Amy — Brand Training 1/12/21.

Hope you have enjoyed your day
Thank you for your hard work.

THATCHERS THEN & NOW:

THE STORY OF A CIDERMAKING FAMILY

James Russell & Neil Phillips

BRISTOL BOOKS

1 Lyons Court, Long Ashton Business Park, Yanley Lane, Long Ashton, Bristol BS41 9LB

Thatchers Then & Now by James Russell

Published by Bristol Books 2014

ISBN 978-1-909446-03-8

Design: Joe Burt (joe@wildsparkdesign.com)

A CIP record for this book is available from the British Library

ACKNOWLEDGEMENTS

I am indebted to Martin Thatcher for commissioning this book and for making my job so much easier by talking so brightly and openly about the company. Many thanks also to John Thatcher for a sparkling interview, and to Ann Thatcher, who kindly dug out photos and shared her memories. I would also like to thank Penny Adair, Maria Forbes, Gary Delafield and the rest of the Thatchers cidermaking team.

This book has been a team effort, and as ever it has been a great pleasure to work with publisher Richard Jones and designer Joe Burt. Lastly, I'd like to thank Neil Phillips, whose photographs make this book a thing of beauty.

P10 SECTION ONE:

THE THATCHERS STORY

P54 SECTION TWO:

TWENTY-FIRST CENTURY CIDER COMPANY

P100 SECTION THREE:

APPLES & ORCHARDS

PROLOGUE BY JOHN & MARTIN THATCHER

This book has been well over a century in the making. Previous generations of our family have all played their part in the Somerset cidermaking story and we are truly privileged to be able to pull every strand together today to create what is not just a story of our family, but a social history of cidermaking in this beautiful county, the home of cider.

We would like to express our sincere thanks to James Russell for his great skill in writing this book, to Maria Forbes for the in-depth study and research she has carried out into our family history over recent years, and to Neil Phillips who continues to capture life at Myrtle Farm through the lens of his camera in a most unique way. Our appreciation goes to everyone who has worked with and for Thatchers Cider since 1904, who have helped create the firm we are today, and finally to the previous generations of the Thatcher family, without whom this book would not have been possible.

INTRODUCTION BY JAMES RUSSELL

Every day thousands of people drink Thatchers cider, in pubs, restaurants and homes across the UK and as far afield as Australia. Cider lovers everywhere will vouch for the quality of Thatchers, but what is it that makes Gold, Katy and all those other ciders so good? As a cidermaker myself (though on a very small scale!) I was fascinated by this question, and so the idea for this book was born. I quickly realised that Thatchers' success in the 21st century is the result of many years' hard work and the gradual build-up of experience over not one or two, or even three generations, but at least five.

Behind Martin Thatcher stand his father John, grandfather Stan, great-grandfather William and great-great-grandfather Benjamin. With the support of their wives, whose contribution should not be underestimated, these men honed recipes, explored new technology, developed networks of suppliers and customers and passed on their skills and experience to the next generation. For more than a hundred years Thatchers cider has been made in the same place, Myrtle Farm in Sandford, North Somerset, by the same family. This is their story.

SECTION ONE

The Thatchers story

Thatchers has been making cider at Myrtle Farm since 1904.

In March 1878 an advertisement appeared in a Bristol newspaper, offering cider for sale in a north Somerset village about ten miles away. Now, generally speaking, traditional farm cidermakers had no need to advertise. Whatever their labourers didn't drink was offered for sale or barter to local people – and the countryside was more heavily populated (and thirstier) than it is today. Yet this cidermaker, whose name was Thatcher, advertised his wares in Bristol, suggesting that he was looking at cider in a different way.

We have no other evidence for his activities but he must have been looking beyond the farm and beyond the village towards a new market in the city. Whether or not he succeeded in selling his wares we have no idea, but in this one modest notice we see the beginnings of what is now a business not only of regional or national, but of international scope.

There have been Thatchers in northern Somerset for more than two hundred years. Thanks to wide-ranging research carried out by historian Maria Forbes, we know first of all that Isaac Thatcher was born in 1806 in Charterhouse, a hamlet on the windswept plateau of the Mendip Hills, the eighth of nine children; he was still there in 1841 when the Census listed him as a labourer who had married servant Harriet Hill a few years earlier. Both signed with an X.

THE FIRST THATCHERS CIDER

While other members of the family scattered around the area, in places like Cheddar and Burrington, Isaac and Harriet remained in Charterhouse – once a major Roman mining centre but by the nineteenth century a settlement of no more than twenty houses, surrounded by old mine workings – where they had four children. Their second-born, Benjamin, was baptized in 1841, and in 1865 married a village girl, Mary Ann Williams. He was an agricultural labourer, as was her father, yet things were about to change, because the 1871 Census tells us that Benjamin and Mary had moved down from the Mendips to Upper Langford, close to Churchill, and that Benjamin was a 'farmer of 6 acres' whose household included a labourer (his brother) and a servant.

According to Maria Forbes, it was he who advertised in the Bristol Mercury, on 29 March 1878:

> *CIDER – Prime new Cider at 30 shillings per hogshead. Apply to Benjamin Thatcher, Upper Langford, Bristol.*

Given the time of year it must indeed have been new cider, fermented but not yet mature and (maybe) too young to drink, at least by modern standards. Was the idea that the buyer would keep the barrel a while before drinking or selling the cider in the summer? Or were tastes different then? Either way, the advertisement confirms what the 1871 Census

suggests: that Benjamin and Mary had the best of their opportunities. That they succeeded is reflected in the fact that, on his death in 1894, Benjamin Thatcher was described as a yeoman, that is to say a landowner on a small scale – a considerable step up from agricultural labourer in the stratified world of the Victorian countryside.

Mary Ann and Benjamin also had four children, and our attention here focuses on William John Thatcher, the youngest, who was born in 1880. William was raised initially in Upper Langford, but after Benjamin's death his mother remarried and moved with her new husband, farmer Thomas Brookman to Sandford Farm, Sandford; William went too, being described in the 1901 Census as 'a farmer's son.' He had begun training as a carpenter before this, but according to his grandson John he had been put off his chosen trade by a nasty incident. In those days, the story goes, a village carpenter doubled as an undertaker, and on one occasion William was instructed to build a coffin for a suicide, and prepare the mutilated body for burial. That was enough for him, and he returned to the family business of farming.

Thomas Brookman was a property owner and a man of some influence, and it was perhaps thanks to his connections that William found himself in 1903 engaged to Beatrice Avery, whose family had been farmers of substance for several generations. Beatrice had been to a Ladies School in Clevedon, and one can imagine that she was quite a catch. A year later William and Beatrice rented Myrtle Farm, a small farm just down the road from Sandford Farm, and so the Thatchers story really begins.

MYRTLE FARM

The village of Sandford has not troubled historians much over the centuries. A cluster of farms attached to the parish of Winscombe, it does not appear in its own right in the Domesday Book. However, Winscombe belonged to the Abbey of Glastonbury, and when 'Sampforde' does appear in the records after the Reformation we find it owned by another major ecclesiastical power, the Dean and Chapter of Wells.

According to Maria Forbes' research, there was a building on the site of Myrtle Farm when Henry VIII was still on the throne. In 1557 William Hawkins bought the lease, and the property stayed in the same family until, some two centuries later, Rebecca Hawkins married into the Brookman family, taking the property with her.

Judging from its thick walls and rough-hewn feel, the oldest part of the farmhouse lying parallel to the road probably dates from some point in the Hawkins' tenure; it is not uncommon to find well-built farmhouses and cottages that have survived since the 17th century. The iron fittings on one set of windows look as they were hammered into shape by the local blacksmith, but otherwise there are few clues. It does seem likely though that the present house was

A timeless scene: autumn at Myrtle Farm.

built on the site of a much older building, while the surrounding fields retain the same essential shape now that they had half a millennium ago.

On a map of 1885 Myrtle Farm is shown with its present L-shape, suggesting that it had been extended sometime earlier. According to the will of Fanny Brookman, who died in 1866, the property included 'stables, coach house, outbuildings, yards, bartons, and gardens and also the close of pasture called The Home Ground and the orchard adjoining and also the orchard opposite…'.

Far from being isolated, these orchards formed part of a band that stretched along the northern base of the Mendips, between the thick woodland covering the hills and the lowlands' rich pastures. In village after village, orchards outnumbered houses and had done for at least a century. When John Billingsley, the great chronicler of Somerset agriculture, visited the area at the very end of the eighteenth century he found that this strip of land produced some of the finest cider apples in the county, the most celebrated of which, the Court of Wick Pippin, was unusual in that it doubled as an eating and a cider apple.

Billingsley reported that the fruit of this region produced 'a cider strong, palatable and highly esteemed as a wholesome table liquor.'

The fact that Benjamin Thatcher was marketing his cider in Bristol suggests that in his day this corner of North Somerset still enjoyed a healthy reputation for the quality of its cider – and, we would have to guess – its apples. But the old traditional orchards were not only planted to produce fruit, but also doubled as pasture. A yeoman farmer wanted his orchards close to the farmhouse, in part so that his few cows would be nearby for milking also because transporting apples any distance on bad roads was a nuisance. Trees tended to be planted as close to the press as possible, and on some old farms you can still see where the lane linking the orchard to the farm buildings has been surfaced with stone, enabling heavily-laden carts to travel with relative ease in wet autumn weather.

It's interesting to think that many Somerset villages owe their origins to apple growing. Where apples prospered, people lived.

The same 1885 map shows that Sandford was a village of small farms and orchards. This country at the northern edge of the hills is fertile, the climate mild thanks to the proximity of the Bristol Channel and the hills themselves taking the brunt of south-westerly storms. The Mendips have been quarried and mined since Roman times, and their populous villages no doubt encouraged the growth of farms in the neighbouring lowlands – and the making of cider.

THE CHEDDAR VALLEY LINE

The opening of the Cheddar Valley railway line in 1876 added a further incentive for local farmers. Running right across the Mendips, it connected Cheddar to the main London-Exeter line at Yatton, enabling people and goods to travel between north

For hundreds of years cider apples have been grown in North Somerset orchards.

Somerset and the capital with an ease that had been impossible to imagine only a few years before. Before long the line became known as the Strawberry Line on account of the huge volume of fruit leaving stations at Cheddar, Axbridge, Draycott, Lodge Hill, Winscombe and Sandford every summer.

Industry used the line too. From Sandford Quarry, on the hillside above the village, a rudimentary railway spur was built down to the station, and for the next fifty years wagons left laden with stone, some of it used to build the new port at Avonmouth. Coal and other goods arrived too, encouraging a haulage company to set up in the village.

So the village where William and Beatrice set up home was connected to the wider world, thanks to the rail network and proximity to the pleasures of Weston, but life there was simple. A local author called Theodore Compton had this to say about Winscombe at the end of the nineteenth century: 'Though it boasts a 'street' and even a *square*, there is not a row of houses in the actual village of Winscombe. The street is without shop, footpath or lamp-post and the square is simply a meeting of lanes. There is not even a public house…'

And this was a hub of activity compared to Sandford, which had a village school and a pub, the Railway Inn, a sub-post office and a shop called The Stores. Local nurseryman Francis Rich specialised in growing fruit trees, otherwise most of the small population were farmers or market gardeners. Only fifty acres in size, Myrtle Farm itself was too small to be named on the maps yet it was, to judge from the description above, a substantial property, with its own coach house. There was a dovecote too, as we shall see, which suggests that the Brookman family had money to invest in building.

A DEEPLY-INGRAINED TRADITION

In 1904, therefore, William and Beatrice became tenants of a small but well-maintained farm, where they raised livestock for the local market, while William continued the cidermaking tradition established by his father. This was perfectly normal. After all there is barely a farm in Somerset that does not have the remains of a cider press built into its walls somewhere, and rare was the farmer who did not put his labourers to work in the autumn gathering apples and pressing the juice. Since the days of the great monasteries, cider apples had been grown in Somerset, many of them varieties found growing wild in the county and adopted for use on account of their particular suitability for cider (we'll come back to these qualities later). And every year for centuries the apples had been gathered and pressed.

As long ago as 1230 Bishop Jocelin of Bath left cider presses in his will, and over the centuries that followed travellers brought descriptions of Somerset cidermaking back from the wild West Country. Around 1700 the traveller and author Celia Fiennes wrote of the very large presses used in Somerset, on

A vintage cider press: layers of apple pulp are made into 'cheeses' and then pressed.

BRISTOL

The raw material of cider: vintage cider apples.

which two hogsheads of juice could be squeezed out of a single cheese – a hogshead being around fifty gallons. To make a cheese, she wrote, 'they pound their apples, then lay fresh straw on the press, and on that lay a good amount of the pulp of the apples, then they turn in the ends of the straw, cover it all round, and lay fresh straw, and then more apples up to the top.'

A hundred years later, John Billingsley, described cidermaking thus. 'The apples are suffered to fall off the trees or when thoroughly ripe are picked with great care. They are then put in heaps to ferment, and remain in that state for three of four weeks; after they are ground and the liquor is expressed…'

This was what basically happened on farms across Somerset every autumn. Cider suited farmers well for many reasons. Unlike beer it requires no heating, and so was and is very cheap to make on a small scale; likewise, once planted an orchard of standard trees can provide apples every autumn for decades – a harvest, as farmers used to say, without a seedtime. Standard orchards provided pasture, shelter and the occasional apple for livestock – pigs and cattle love apples – while the animals provided nutrients for the trees in the form of manure.

Moreover, the annual cycle of cidermaking fits easily into the yearly round of the farm, with the apples ready for harvest after most other crops and pressing taking place late in the year when there was little else to do. Pruning, tree-planting and other orchard jobs also need to be done in the depths of winter, when the farm is otherwise quiet.

Before the arrival of tractors and combine harvesters, when harvest was still a great communal undertaking, cider was consumed in huge quantities by men, women and children – up to two gallons a day for the strongest men. By the middle of the 19th century it was generally accepted that workers were being paid around 15% of their wages in cider – a payment-in-kind known as Cider Truck – and by the end of Victoria's reign horrified MPs in Westminster had passed laws against the practice. Then, West Country farmers were not all that interested in what Parliament thought they ought or ought not be doing, and they carried on much as before. Even in the 1930s, men drank great jars of cider as they made hay and took in the harvest, and at the end of the day the farmer was expected to provide a hearty supper of bread and cheese in the farmhouse kitchen, washed down with lots more cider, followed by music and song.

WILLIAM THATCHER'S CHAMPAGNE CIDER

William Thatcher made cider partly to give the men who worked for him, particularly at harvest time, and his recipe, made using a blend of old-fashioned Somerset cider apples (of which, more below), forms the basis for today's Thatchers Traditional and Vintage ciders. But William did not only make draught farmhouse cider; he also made a very different drink – the West Country's version of champagne.

Actually this isn't quite the right way round: although French readers might dispute this, the in-bottle fermentation process now legally called the Méthode Champenoise was invented on *this* side of the English Channel, by 17th-century Herefordshire cidermakers. Charles II wasn't called the Merrie Monarch for nothing, and he and his courtiers drank a strong, sparkling cider that was kept in heavy glass bottles (a new invention) with wired-on corks, which was clearly bottle-fermented.

In Somerset too it was common for producers at that time to bottle cider with a few raisins or a small piece of sugar added, to give it sweetness and sparkle. They understood racking and were able to control fermentations to create soft, sweet ciders; the cidermakers, John Billingsley observed in 1800, racked their cider after only a few days' fermentation, then repeatedly thereafter. Interestingly, he described them not as farmers but as gentlemen. There was a second cider tradition in Somerset working alongside the farm presses, that of landowners and clergymen – most famously, Parson Thomas Cornish of Heathfield Rectory, Taunton, who sold cider to Queen Victoria – who made strong, palatable cider and drank it instead of wine.

In this country, at least, cider and perry are unique in being made and enjoyed both as an inexpensive and refreshing draught drink and as a sophisticated tipple. William Thatcher understood this, and he made both farmhouse scrumpy and bottle-fermented champagne cider, the latter perhaps only for the enjoyment of family and friends.

STUDYING CIDER AT LONG ASHTON

He may have been inspired partly by events unfolding a few miles to the north of Sandford, in the village of Long Ashton. There, in 1904 – the year that William first made cider at Myrtle Farm – Britain's first and only institute of cider science began pressing apples. There had been concern in the previous decades that West Country cidermaking had lost some of its zip and zing, and in the 1870s Somerset landowner Robert Neville-Grenville began privately researching the quality and characteristics of different cider apples. From his efforts came the foundation of the National Fruit and Cider Institute, which started work in a modest way in 1903, being renamed the Long Ashton Research Station (LARS) a few years later.

LARS galvanised the local cidermaking community. Producers were invited to submit apples for analysis and identification, and given advice on the best apples to grow. Terms such as 'bittersweet' appeared for the first time, as we shall see, as cidermaking was put on a scientific footing for the first time. Cider was also legally defined for the first time, when a group of Somerset producers went to the government's Board of Agriculture and requested the prosecution of two firms for selling, as cider, liquor

Traditional thatching at Williams Barn, Myrtle Farm.

Thatchers have pioneered the planting of hedgerow orchards, bringing an ancient tradition up to date.

that did not contain apple juice. The prosecutions were successful and the principle enshrined in law that cider must contain a certain proportion of apple juice. Meanwhile, an annual cider competition was established at Long Ashton, and for the next twenty or thirty years, Somerset cidermakers vied to produce the best cider. Along with other techniques, the champagne method enjoyed a revival, with Clapps of Baltonborough advertising 'Avalagne – Champagne of Avalon'. Perhaps William Thatcher hoped to produce a competition-winner, or perhaps he was simply a man of taste. Either way, the fact that he went through the time-consuming and difficult process suggests that, while he only made cider on a small scale, he took his work seriously.

STAN THATCHER: FARM CIDERMAKER

William and Beatrice had five children, of whom one, James, died in infancy. A daughter, Winifred, died in her early 30s, leaving Stanley, Reginald and Betty. The eldest son, Stan took over Myrtle Farm when his father died in 1937, while Reg first went to college then took a job with Bristol Sand and Gravel, before moving to Chelmscombe Quarry and eventually owning a prosperous construction business. Of the two brothers, Reg had the commercial acumen, being good with money and having his eye firmly on his business; if he was out on a Sunday drive he'd pick a route that allowed him to price up a job while he was at it.

Stan Thatcher, with barrels.

Stan was no businessman, being too good-natured to chase debtors for money, but his wife Molly, whom he married in 1940, was made of sterner stuff. Her father was a shrewd businessman, the type who knows exactly what he has in the bank without ever

writing anything down, and it fell to her to keep the farm finances in order. It was Molly, incidentally, who kept a stick in the corner of the kitchen and wielded it occasionally over unruly boys, whereas Stan never raised his hand or, for that matter, his voice. Their son John was born in 1943.

World War II brought change to Sandford as to other villages, with land girls replacing the men who had gone off to fight and evacuees fleeing Bristol. Winscombe and district were spared more than the occasional enemy bomb dropped by accident or to lighten an aircraft (there were bomb craters in Webbers Orchard from jettisoned bombs), but everyone witnessed the ghastly spectacle of the sky lit red by Bristol burning during the Blitz of November 1940. Other changes had already begun before the war, as more and more cars appeared on the roads – the first petrol garage was established in Sandford in the mid-1930s, and a road contractor set up shop around the same time. As people began commuting to Bristol in large numbers the population of Winscombe, Sandford and the surrounding villages grew markedly, and this growth accelerated as peace returned.

In many ways, however, life on a small mixed farm was much the same as it always had been. Every morning and evening Stan and Molly went across the road to the orchard opposite (the same one mentioned in Fanny Brookman's will almost a century earlier), where they each milked six or seven cows by hand, the milk then being collected from a stand at the farm gate. They kept sheep, pigs and chickens, and grew crops for feed, and in amongst all these activities made cider in season.

A thriving bee and insect population is essential for pollination, leading to a good crop of cider apples.

Stan with grandsons, Martin (r) and Robert (l).

CIDERMAKING IN THE OLD WAY

The cidermaking itself was equally timeless. Apples were allowed to fall from the trees and left to lie in the orchards until November, then picked up by hand, collected in sacks, milled with a scratter and then pressed in the old manner, the pulp wrapped in layers on the press and juice collected as the cidermakers used sheer physical strength to turn a handle and increase the pressure from above. Labour-intensive but otherwise cheap and straightforward, this method is still used by many cidermakers, but the kit imposes a physical limit on the volume of apples that can be pressed; using this method Stan was able to press two tons of fruit a day, one between milkings and one in the evening.

Unlike his father, Stan concentrated on making old-fashioned draught farmhouse cider, made with a blend of bittersweet apples to give it that distinctive flavour and, just as important, mouth-feel. This quality, peculiar to cider produced in the West Country, is immediately obvious when you take a sip, but difficult to describe well. The best bet is to have a glass of Thatchers Traditional, and all will be revealed, since Traditional is – as the name suggests – still made to Stan's recipe.

Although no new orchards were planted at Myrtle Farm in Stan's time, the press was kept well supplied with fruit brought in from orchards in the vicinity, the owners of which were no doubt happy to take payment in the form of finished cider. Every village and every farm had orchards of standard (ie full-sized) apple trees, which may have been planted fifty or more years before, so why go through the extremely laborious process of planting new orchards?

Besides, Stan enjoyed the social world of cidermaking, with the coming and going of people with sacks of apples for the press. But the farm had to come first, especially during wartime when every foot of land and ounce of energy was devoted to food production, and the business grew slowly. In 1951 or thereabouts came the first major change, when Stan installed a hydraulic press that was faster and more efficient than the old one. Now he was able to press about six tons of apples a day, and so the situation remained for almost twenty years, until father and son decided that John should take over the day-to-day running of the farm, while Stan concentrated on the cider business.

A CIDER SHOP... IN A DOVECOTE?

The press had its home in a building close to the farmhouse, while scores of wooden barrels stood in rows across the yard; morning and evening the cows passed them on their way to the milking shed (put up when the farm expanded after the war). As for Stan's shop, it occupied a small building that still stands at the entrance to Myrtle Farm – the old dovecote.

Once fairly common on farms and country estates, dovecotes housed pigeons which were kept for their meat and eggs, and also for their droppings – which were a valuable fertilizer. But changing tastes saw farmers hunting birds rather than keeping them, and by the 20th century dovecotes across the country were being put to new uses. A former dovecote made an excellent cider house, or cellar as it was known at Myrtle Farm, so it was here that Stan sat on a barrel fitted with a circular foam cushion and served cider from the barrel to his numerous local customers.

This cellar was not a shop in the modern sense,

Paul Lilley pouring cider from the barrel, circa 1990.

Low tech: Stan Thatcher filling containers from the barrel.

nor was it a pub. It was a place where people came to buy cider, true, but rare was the customer who would walk in, make a purchase and leave. On entering, they would be asked what they cared for, the choices being dry, medium or sweet, and Stan would fill a pint glass from the collection wedged into the niches once occupied by pigeons, and hand it over. No money exchanged hands at this point. The first pint was meant as a taster, though you would imagine that most of Stan's customers had an inkling of what his cider tasted like, and it wasn't uncommon for people to end up drinking far more cider 'on the house' than they paid for.

Really it was a social custom, since the cellar doubled as a talking shop for farmers, local people and others of a like mind – a place to exchange news and gossip in an informal and congenial atmosphere. No cider house of this kind was ever bought by a brewery; these were places where friends gathered, and money played only a minor role in what was essentially a social business. Many were the happy customers who went wobbling off on their bike or, as was the norm in those days, weaving off down Station Road in their car, having enjoyed Stan's company for an hour or three.

There were not many frills. In fact there were none. Stan kept a container full of water in a corner, and when a customer had finished with his glass it was sloshed about in said container before being replaced in its niche, ready for the next customer. There were plastic containers and glass jars which Stan filled with cider for people to buy, but otherwise the scene would have been familiar to a visitor from fifty, a hundred, even two hundred years before. This was a cider business that had one foot in the present, and one foot in the past, and thoughts of the future only extended as far as the next autumn's apple crop. The best cider, Stan liked to say, was made on 1st November, and drunk on 1st June. Nobody could argue with that.

Change really began in the early 1970s, when Thatchers bought three Fielding and Platt hydraulic presses from another cidermaker. Now five hardworking men could press anything up to forty tons of apples a day, a huge increase that inevitably brought with it further changes. New markets had to be found for all the extra cider, and by the middle of the decade Thatchers cider was being sold in pubs both in local village pubs and further afield. It helped that the population in the surrounding villages had been growing steadily since the war, with Winscombe already approaching the size of a small town. The Cheddar Valley Line closed as more people took to the roads – which meant more potential customers passing Myrtle Farm every day.

FROM FARM TO BUSINESS

Like his father, John was born at Myrtle Farm and as he grew up learnt every aspect of running a farm that had increased substantially in acreage but was still a small, mixed outfit; in those days cowsheds covered the northern side of the yard and the cattle were brought in along the road for milking twice a day. There were pigs to look after, and sheep, as well as crops grown mostly for animal feed.

By his own admission John is more of a businessman than his father was, and in the 1970s, as he became experienced at running the farm, he began to look at it with a businessman's eye and perceived a need to change. He quite rightly saw that the farmer was at a disadvantage economically, because he sold produce – milk, or livestock, or crops – at a price he was unable to control, yet was at the mercy of market forces when it came to buying feed or fertilizer or diesel. Pigs were his biggest concern, with 150 sows at the farm at one time, but even finishing pigs for pork he found himself unable to profit fully from the enterprise. There was no way of branding his products, of making a name for quality and so building up the business.

The answer, he saw, lay in moving a step closer to the consumer, and to this end he took the bold step of buying a wholesale fruit and vegetable business from a company in Weston-super-Mare. This may not seem so different from farming but in reality it was a major departure, since he was able for the first time to seek out new markets for his produce – now, with lettuces, bagged up and ready for the shelf as 'products'. Significantly, this meant not only that John had to sell his products to customers, but also that he was able for the first time to develop a brand, something farmers have not traditionally done.

As well as his father's support, John enjoyed the wise counsel and support of Ann, whom he had married in 1966. Raised on her parents' farm in Upton-on-Severn, Ann had always dreamed of being involved in medicine, and after leaving school had come to Bristol to train as a radiographer. Through family connections she found a room in the village of Star, only a few miles from Sandford, and while staying there went along to a function organised by the local Young Farmers. So she met John, and after their marriage the couple embarked on an adventurous life, both business-wise and later as travellers to many countries of the world. They had two sons, Martin (who we will hear a lot more about) and Robert, who runs the farm in Upton-on-Severn where Ann grew up.

By the 1960s the hundreds of standard orchards along the northern base of the Mendips had grown too old to be productive, and with urgent demand for housing they were sold off one by one. Stan himself sold the old orchard across the road from Myrtle Farm to a builder, and when John and Ann were married they first lived in a house on Orchard Drive, as the new development was called.

But when the boys came along they swapped over, so that Martin and Robert were raised, as their father and grandfather had been, on the farm, where both family and business life revolved around the farmhouse with its solid walls, antique beams and long table. As at many farms the kitchen table was the setting for important meetings, where sometimes life-changing decisions were made, and there were always piles of paperwork in corners and a constant coming and going of people. In the early days customers for the cider house would sometimes tap at the window, and on Saturday lunchtimes the farm's small team of workers came to the kitchen window, to be given their pay packet by Ann from a small, circular tea tray.

HEALTH AND SAFETY OFFICERS, LOOK AWAY...

Like their father, Martin and Robert grew up on a busy working farm, but one that was changing quickly. By the early 1970s traffic on the main road was making it difficult to move cows at milking time and the decision was taken to give up one of the staples of the mixed farm. Like so many other decisions made by the family it turned out to be timely, given the difficulties faced by small-scale dairy farmers in the supermarket age. Importantly, the departure of the cows made more sheds available for other uses.

Much of the farmyard was taken up by rows of wooden barrels, hogsheads and pipes (a barrel twice the capacity of a hogshead) in which all of the cider was kept. The building that is now the cider shop was home to the fruit and veg business, while the pressing still took place in the building closest to the farmhouse, with the cider shop continuing to occupy the dovecote. The brothers grew up with their parents and paternal grandparents close at hand. During term-time Stan drove them to school in Weston, with a Polo mint for the journey there in the morning and a Mars Bar for the journey back. When they got home they would jump out of the car and charge around the

This miscanthus will fuel Thatchers' biomass boiler.

farmyard, racing each along the tops of the barrels in a manner that would give a modern Health & Safety officer apoplexy.

TWELVE HUNDRED LETTUCES

Theirs was a typical farm childhood, in the sense that there was little distinction between work, play and social life. As soon as each of them was old enough to take on a particular responsibility it was given to them, and there was always plenty to do. As part of the fruit and veg business, John grew lettuce, cabbage and leeks on a good plot of fertile land along the road from the farm, and one job Martin still remembers with particular clarity involved lettuces. On Saturday mornings during term-time he would be presented with a hundred dozen heads of lettuce, fresh from the field. He then bagged each one up and built cardboard boxes, placing a dozen heads in each before loading them onto a lorry. He probably had help, he admits, but in his memory it's just him… and twelve hundred lettuces.

The offending vegetables had to be taken for sale to the wholesale market in Bristol, where John would also buy produce for the fruit and veg business. For the boys this often meant getting up at four in the morning and travelling into Bristol with their father, loading and unloading while the necessary deals were made, coming home and then – in term-time – going to school. They saw nothing strange in this, as children on other farms did similar things, neither did they think of it particularly as work. It was just part of life, same as the delivery rounds taking produce to the local schools and shops that John had won as customers. When 25kg bags of potatoes needed delivering to chip shops, Martin or Robert would go along, their job being to drag the sacks to the side of the lorry so the deliveryman didn't have to climb up.

There were animals to look after as well. For the cattle Martin worked on what they called the 'mill and mix', filling a grain silo with barley, then grinding it and adding molasses and vitamins before bagging it up in old Nitran sacks and taking out to the fields. Summer excursions to Salisbury Plain provided bales of straw for bedding, Martin driving off with a couple of the men in the lorry nicknamed KTK after its number plate. John employed no more than half a dozen men, but working with them was an invaluable experience for his son.

At an age when most boys are content to kick a football around Martin was driving a tractor, and only a couple of months after his thirteenth birthday John taught him to plough. A favourite job was driving the tractor between the rows of trees in Webbers Orchard, not only cutting the grass but harvesting it for silage. In the summer holidays he might spend several weeks going up and down, back and forth, enjoying every minute. When he was seven he had helped his father and grandfather plant this first bush orchard – which we will come back to – so Martin grew up alongside these first apple trees, the planting of which was a

The trees awake from their winter slumber.

Hands on: Martin shares his barrel-rolling skills with his son Peter.

major expression of faith in the future of Thatchers Cider.

ROLL OUT THE BARRELS

At this stage apples were still collected by hand, a job the brothers carried out in the old way, picking the fruit up off the ground late in the autumn. Pressing carried on until Christmas Eve, or sometimes into the New Year, and remained a laborious business despite the new gear.

Indeed, the whole cidermaking enterprise was labour intensive, with a host of time-consuming jobs. Photos of Myrtle Farm from the early 1970s show the cowsheds still standing and, in the western corner of the yard, the rows of wooden barrels. Anyone who makes cider will agree that one of the most important rules for success is to keep all equipment spotless, but how do you go about cleaning scores of large wooden barrels – which you can't treat with chemicals – prior to filling them with juice? The answer at Thatchers in the 1970s was to roll each barrel across the yard, to the tap near the farmhouse, and use a hose to part-fill it with water. A length of metal chain was then dropped into the barrel, so that when the barrel was rolled around the chain scoured the inside - an ingenious answer to the problem but one which still left someone with a solid week's work.

Not that there's anything wrong with hard work. Farming had always involved constant, year-round toil, but it was satisfying work carried on among familiar people, in a familiar place. Martin's recollections of his boyhood suggest that he was constantly busy but – if we discount bagging lettuces – enjoying what he did and relishing the responsibility of driving mowers and tractors and, later, delivery vans. He and Robert also rode ponies and played sports, but it was the day-to-day farming life that was in their blood, and which they both loved most.

In 1981, at age 13, Martin went to board at Wellington School, and only a few weeks later received the sad news that Stan had died at the age of 71. An era had come to an end with the old man's passing, and many would miss him. Indeed, such a crowd turned up to pay their respects that there were more people outside the packed church than inside.

CIDER TAKES OFF

After Stan's death John continued as before for a while, but he became increasingly aware of the need to simplify, to focus. If he had been living fifty years earlier he might have been perfectly happy running a small mixed farm as his father and grandfather had done, but times were changing. The supermarket age was underway and farmers were being encouraged to 'diversify'. But where did the future lie in this revolutionary decade? Of all the various activities going on at the farm, where was the opportunity for growth, for investment? With food prices being steadily forced down by ever-cheaper imports, the raising of livestock was becoming less profitable. The future of the fruit and vegetable business did not seem too rosy either. Out of everything, in fact, cider looked like the best bet for the future.

Not that this conclusion was reached easily, since the cider industry was in a state of some turmoil and had been for a while. By the late 1960s many of Somerset's farm cidermakers had abandoned their presses, as demand for cider fell with the replacement of hard-working labourers by combine harvesters and other machines. The days of Cider Truck were nearly over. Smaller producers were also hit hardest by the decision of Chancellor Dennis Healey, in 1976, to levy duty on cider.

This had happened before, most famously in the late 18th century when there were such protests that repeal soon followed. On this occasion, though, there was an air of permanence about it – of things being brought properly into line. While the very smallest producers, selling fifteen hundred gallons or less a year, were exempt, everyone else lived from then on in the anxious knowledge that every budget might bring a rise in duty.

Yet there was also plenty of cause for optimism. After all, Mr Healey would hardly have bothered taxing cider if the industry wasn't in robust good health, and in the Bristol region there was no shortage of demand. With a hot summer or two, the Wurzels at the top of the pop charts, and TV sitcom The Good Life inspiring people to embrace simpler things, farmhouse cider was suddenly fashionable. Like real ale, which was also on the up, it offered people authentic drinking pleasure far removed from unpalatable tinned lager. For a cidermaker seeking to expand there were opportunities to be found. Pubs were difficult to deal with, as they had been for many years, because so many were tied to breweries that had contracts with national cidermakers, but the Bristol area offered plenty of scope for selling barrels of draught cider to off licences – as Benjamin Thatcher had once done.

CHEDDAR VALLEY COMES TO MYRTLE FARM

Then came a surprise opportunity. For many years Peter Champeney had been making cider just down the road near Sandford Station, his Cheddar Valley brand a regional favourite renowned for its unusual carroty orange hue. But in 1983 he decided to call it a day, and in one of those leap-of-faith moments, John bought the business, which included the land, the brand and all the equipment. And Cheddar Valley was no small-time craft cider. Over the years Peter Champeney had collected together a formidable collection of enormous oak vats – each as tall as a house – plus he had three flat-bed presses, just as Thatchers had.

John moved them to Myrtle Farm, thereby increasing the company's pressing capacity. By now the rows of wooden barrels had been replaced with black Rotoplas fermentation tanks, each of which held

Abundant spring blossom promises a good apple crop.

330 gallons (as opposed to the 40-100 gallon capacity of the barrels), and soon after they were joined by a pair of stainless steel tanks bought from a winery and half-buried to reduce their intrusiveness in the village skyline. These arrived full of small plastic balls that were designed to keep air out of the wine as the tank was gradually emptied; naturally the boys had the job of fishing them all out. With the addition of the new presses, production doubled to two hundred thousand gallons a year, lifting Thatchers into the big league, at least in Somerset terms.

Good old farmhouse scrumpy.

FARMER'S TIPPLE

The 1980s offered challenges and opportunities. A rather unfair challenge was posed by the family name, which was very well-known and not very popular in certain parts of the country thanks to the politics of the day and one Margaret Thatcher. So it was that a wholesaler which sold cider in South Wales and the Midlands made a request to market Thatchers cider under its own brand names of Farmer's Tipple and Old Mill, which it did successfully for some years.

Opportunity, meanwhile came with the growing popularity of music festivals. Draught cider was the drink of choice, being the perfect drink for a summer's day in the country (and affordable with it), and on several occasions Thatchers were asked to provide large volumes of cider for a particular festival in the Midlands. The first order was one of the most impressive the company had received and inevitably it had to be fulfilled in a matter of days.

It was for five thousand gallons of cider, in one-gallon containers, and the prospect of filling these with only a simple hose was daunting to say the least. That the order was completed on time and with the minimum of fuss is testament to John and Martin's ingenuity. The farmyard had a gulley running across it, used for drainage, which they realised was just the right width to accommodate a plastic gallon container; it would also stop cider spilling all over the yard. So the containers were lined up and cider siphoned by hand straight from the tank, then they were loaded onto a tipper lorry hired from Sweetings, and off they went.

As the cider equipment moved in, John and Ann made a conscious decision to wind down other aspects of the farm, not by reducing the overall

The best cider comes from the finest fruit.

activity or by selling any of their three hundred-plus acres, but by simplifying what they did. Raising pigs is complicated and difficult, whereas fattening beef cattle is straightforward, so they gave up the former in favour of the latter. Growing lettuces is complicated, but growing apples is, again, relatively easy. So the vegetable-growing was abandoned and more orchards were planted, so that the Home Ground mentioned in Fanny Brookman's will was soon covered in apple trees.

Not that these changes were irreversible; the trees were put in to provide a crop, not as a permanent addition to the landscape, and the land could easily have been put back to grass if necessary. There was always a Plan B with John, who perhaps knew too much about both farming and the cider industry to assume that life would always run smoothly. After all 1984 had seen another aggressive hike in cider tax duty, and five years later Devon lost its best-known cider company when Whiteways closed down.

ABOUT A BUCHER

At Myrtle Farm, however, growth continued inexorably. The same year Whiteways closed, John made his biggest commitment yet to the cider business. Commitments, in fact. First came the decision to close the old Cheddar Valley works down the road and sell off the land (which today is home to a retirement village run by St Monica's Trust). This meant bringing the bottling equipment that had been set up there to Myrtle Farm and also, and most importantly, the giant oak vats. A team of coopers was brought in, and they took apart the oversized barrels, getting covered head to foot in thick black dust – the residue of decades – as they did so. The great planks and iron hoops were taken by lorry up to Myrtle Farm and there reassembled, and such was the coopers' skill that they look today as if they have never been moved an inch since they were first built.

The second of that year's great changes was the acquisition of a Bucher HP5000 press from Merrydown Cider in Sussex. Martin remembers this particularly well because his 21st birthday party ended at 4am on a Sunday morning, and at 4am the following morning he was sitting behind the wheel of a lorry, headed east, with a long and potentially trying day ahead.

For the benefit of anyone who isn't an expert in fruit pressing technology, a Bucher press is a quite different animal to the kind of flat bed press traditionally used on Somerset farms. It isn't anything like a modern belt press either – more like a giant syringe. Apple pulp is pumped into a cylindrical chamber through which a number of flexible tubes run lengthways, then a piston compresses the pulp, squeezing the juice through holes in these tubes. The chamber can be rotated or more pulp added, and the process repeated until every ounce of juice has been squeezed out of the pomace – which is ejected automatically and spirited away on a conveyor.

The orchards of Myrtle Farm are a haven for wildlife.

Start up your Bucher and almost all of the manual labour associated with pressing apples disappears. This press can be operated by one person, and they don't even have to be particularly strong. It is completely sealed off from the outside world, so there is no chance of the pulp or juice becoming contaminated, and it is devilishly efficient, automatically applying just the right amount of pressure for the type of fruit being pressed – because a Bucher HP5000 is a universal press, and you'd struggle to come up with a fruit it can't process.

There are, however, a couple of downsides to this miraculous machine. First, the price tag. In 1989 £25,000 was a lot of money to shell out on a second-hand press. Second, its size. When the Thatchers team drove over to Sussex to pick up the press they took a hired flatbed lorry for the press and a second lorry for fixtures and fittings. A crane was on hand for loading purposes, and a day set aside for the job, because the press alone weighs twenty-five tons. Given the scale of the operation, the complexity of the new kit and the ever-present possibility of something going wrong, the company budgeted a further £25,000 for the move and installation, a budget they managed to stay within thanks to some typically canny problem-solving during the set-up.

Interestingly, the new arrival didn't increase the amount of fruit John was able to press every year, but it massively simplified the operation. Gone were the half-dozen flat-bed presses and all the labour associated with them – the building of the cheese and removal of the pomace, the constant lifting and carrying. Cider had always been cheap to make but incredibly hard work. The new press didn't necessarily make the whole process of crafting good cider easy, but it did reduce the physical labour involved; suddenly there were hands that could be put to other work, and minds that were free to think. From here, further change was inevitable.

ADVENTURES IN BOTTLING

By the end of the 1980s Myrtle Farm was surrounded by orchards, while the yard itself was covered in sheds housing the equipment necessary for pressing apples, fermenting the juice and packaging the cider.

In the early days this meant pouring it into five-gallon barrels or gallon containers, but as John focused more on the cider business he began exploring the possibilities of bottling. The first experiments were carried out in the building that now houses the shop, and they were fairly basic. Essentially cider was circulated from one of the black storage tanks on a loop, to which was attached a manifold fitted with four squishy rubber pipes. Below these was a wooden tray with a formica bottom. There was room for four bottles on the tray, and the art was to keep all four pipes running, pinching one to stop the flow when the bottle below was full and quickly switching the bottle for an empty one.

Caps were put on by hand, and so the first bottles

A bird's eye view of Thatchers.

of Thatchers cider were sent out into the world.

A time-consuming operation, perhaps, but one that allowed John to explore the methods and techniques of bottling at a basic level. Rather than rush in to buy new equipment, he started simply, using makeshift equipment that was close to hand, and only gradually, after much thought and discussion, moving up through the gears.

The next development in bottling took place at the former Cheddar Valley works, where John and his team set up a system for filling what are known in the trade as PET (Polyethylene terephthalate) containers or to the rest of us simply as plastic bottles. This was more elaborate than anything that had gone before, principally because steps now had to be taken to preserve the cider.

The tannins and acids in traditional unfiltered cider act as a preservative in the short term, but there have always been hazards involved in storage, from acetification (cider turning to vinegar) to microbial infection. Even when nothing goes wrong you often find variation from barrel to barrel in a single year's pressing. One might be cloudy, for example, the next clear.

Drinkers of traditional cider tend to be aware of this and even relish slight variations, but once cider is bottled and put on a shelf different standards begin to apply. Besides, nobody wants a customer to uncap a bottle of cider and find it sour, so steps have to be taken to prevent any changes taking place in the cider, which means either putting it through a fine filter to remove any remaining yeast and nutrients, or pasteurising it (heating the liquid to achieve the same result). Today Thatchers uses the best available filtering system, but back in the 1980s John invested in a pasteurising machine capable of processing 120 gallons an hour. He also bought a carbonator, so that often pasteurizing would take place on a Monday, cooling through Tuesday and bottling Wednesday. This might seem slow today, but compared to filling four bottles at a time by hand it was fantastic.

With the sale of the Cheddar Valley site, bottling moved back to Myrtle Farm and became more fantastic still. Now the normal procedure was to finish bottling at lunchtime, then spend ten hours or so pasteurizing enough cider to fill a clean 1200 gallon tank. This might take until almost midnight, after which the cider was rapidly cooled, ready for bottling the following morning.

So the turn of the decade saw the company producing more cider than ever before, but keeping things simple still. The shop had made a move from the dovecote into the larger building next door, and there were more ciders to choose from: Farmhouse Scrumpy in dry, medium or sweet, along with Cheddar Valley and new brands like Mendip Magic (which was fizzy) and Sandford Superb (which wasn't). Thatchers' lorries travelled the region, delivering cider to an ever-increasing range of shops and pubs, and there were deliveries further afield to wholesalers in the West Midlands and Wales. This was an operation with few frills, producing and delivering top-quality Somerset farmhouse cider.

MARTIN JOINS THE BUSINESS

You may be wondering at this point what the future Managing Director of Thatchers was up to while all this was going on. Martin had left school in 1987, just before his 19th birthday, his heart set on running a farm. All through his schooldays he had spent holidays working at Myrtle Farm, and he had also continued his apprenticeship at his maternal grandfather's farm in Upton-on-Severn, where he learnt how to manage sheep and grow cereals. From the age of 16, indeed, he was up there every summer harvesting with a Claas combine, which his grandfather had taught him to drive.

After leaving school he set to work at Myrtle Farm, spending a season making cider on the flat bed presses and helping John with the livestock. Then came an opportunity. His grandfather wanted to retire from farming and needed someone to rent the farm. Having just turned twenty, Martin jumped at the chance, and for a year commuted between Sandford and Upton (a round trip of some 140 miles), where his apprenticeship was soon almost complete; he had learnt just about everything he needed to know in order to manage his own farm, and only a brave man would have bet on him going into any career but farming.

A heart-warming sight...

One big gap in his knowledge remained though. Despite all his experience he had never worked with dairy cattle so he went off to try his hand at his uncle's dairy farm, but quickly realised that he didn't enjoy milking morning and night. For the first time he began to have doubts about his future as a farmer, and almost on cue John phoned from Sandford: a delivery driver had hurt his back, so could Martin come home and drive the lorry instead? He went. Today it is his brother Robert who farms at Upton-on-Severn...

Until now Martin had been involved only occasionally in the cider business, pressing fruit in the autumn and delivering pomace to local farms. The delivery round was a new experience, taking him across the Mendips, to Bath, and into South Bristol suburbs where life was rather different than in quiet Sandford. Learning rapidly that you needed to have your wits about you when driving around with a

John and Martin in the 1990s, with the latest in packaging.

lorry-load of highly desirable cider, he began to see a future different from the one he had always imagined. He went on a pruning course at Bulmers and spent much of the winter in the orchards, then took to the road for a year driving lorries. Next, he became the sales team and spent another year travelling the region in search of new business, and having now seen what life was like for the people on the front line of the cider business, he was ready to get involved in the management of the company his father had gradually transformed over the previous decade.

The cider industry was going through one of its growth periods in the early 1990s, but most of the brands driving this boom would have been unrecognisable to Stan Thatcher. This was the era of white cider, and bottled ciders with names and flavours that had little to do with apples; often these products seemed to be masquerading as something other than cider, and quality mattered less than affordability.

This wasn't Thatchers' style at all.

INTRODUCING KATY

Cider had always been made with care at Myrtle Farm, but with the arrival of the Bucher press John and his cidermaking team began to think more creatively. In the wine industry it has long been common practice to ferment the juice of different grape varieties separately then blend the wines to taste. This enables winemakers both to match the flavours of previous years and to experiment with new blends, and in general it gives them much greater control over the end product.

So what if you were to do the same with cider? A long tradition of single variety cider exists in Somerset; the early maturing apple Morgan Sweet is traditionally pressed in August to be ready for Christmas as a single variety, and we forgive the lack of body and bite for the treat of new cider in December. In the past, however, few cidermakers relied on a single variety for their cider. Given the variations in cropping and fruit quality common to cider apples, it made more sense to develop a flavoursome blend of tried and tested varieties, which would be broadly similar each year. Thatchers Traditional is such a blend, a descendant of those 'strong, palatable and highly esteemed' ciders which has served the company and its customers well for more than a century.

Neither Stan nor John had the kit or the inclination to press different bittersweets separately and then blend the finished ciders. But when, soon after the new press was installed, a load of Katy apples arrived from Kent, the cidermakers realised they needed to do something else with them.

Cider has always been made with eating apples in the East of England, though ordinarily as a blend of different fruit. What John and Martin decided to do was press Katy separately and see what they got.

Most cider apples taste vile when you bite into

Katy and friends, late 1990s.

A similar shot, but note the subtle rebranding.

them but yield a rich, delicious juice. Katy tastes light and sweet, but would it yield a juice robust enough to make a good cider? The answer was yes. Katy apples produced a cider that was strong in alcohol, but delicate in appearance and flavour. The farmers of the old days would have dismissed it as a Lady's Cider, but times were changing. At first Katy was something of a novelty. There was little in the way of marketing in those days, and not much scope among Thatchers' regular customers – mostly local shops, pubs and off licences – for this new kind of cider. Mendip Magic and Sandford Superb were still the 'modern' brands, but their days were numbered…

MEETING TESCO

By 1993 Martin was fully embroiled in sales and marketing, and that year a new and exciting opportunity presented itself when the company's experienced marketing man Ron Eddy secured a meeting with Ian Parker, a buyer at Tesco. This was an achievement in itself for a guy representing a small-scale producer like Thatchers, and given Martin's minimal experience of the corporate world he was understandably both excited and terrified. Shoes polished and samples of Mendip Magic and Traditional all ready to show he drove off with Ron full of excitement yet, at the same time, not really

Blossom makes apples and apples make barrels of cider.

THATCHERS

believing that anything would come of the meeting. What followed is a scenario that ought to be taught on business courses.

Martin and Ron were called into the buyer's office, where they chatted briefly about Thatchers, what the business did and what it stood for. About five minutes went by, and then Mr Parker spoke. Here's how it is, he told them. If you have brought along products that are just the same as the ones already on the supermarket's shelves, then I'll listen to what you have to say but you won't be invited back and your products won't be listed. If, on the other hand, you have something new and really interesting, and we can agree terms, then you will be.

At this, Ron quickly intervened and asked if they could come back in two weeks with 'something new and really interesting'; the buyer said yes, and after a meeting lasting no more than ten minutes they were heading back to Sandford. As they drove away, Martin wondered out loud what had just happened. Was that it? Had they blown it? No, came the reply. If they'd gone ahead with their plan of offering their existing ciders they would have been politely shown the door. As it was they had a second chance, but what on earth were they going to take along next time?

Here was a fabulous opportunity to launch into a national market but there was no instruction manual telling them how to go about it, no training anyone could have undergone. The existing Thatchers ciders were not suitable, so new ciders had to be chosen, bottled and branded - the sort of operation that would take a big company years, and they had only a few short weeks. But sometimes a car ride gets the thought processes going, and by the time they arrived back at Myrtle Farm they had hatched a plan. If Tesco wanted new and interesting they would get it, in the form of a cider nobody outside the company had seen or tasted before.

The two weeks passed in feverish activity and many kitchen table conferences. It was decided to take Katy, as the absolute novelty, and also single variety Dabinett, Somerset Redstreak and Tremlett's Bitter ciders, the varieties having been pressed separately the previous year. Martin and Ron bought drinks in 500ml glass bottles, which they emptied and washed, soaking off the labels. There were no labels for the new single variety ciders so they mocked some up, filled the bottles and then polished them – with Pledge – until they shone. They even polished the caps.

These gleaming bottles were carefully wrapped in raffia paper and then off the sales team went for meeting number two. This one was, if anything, shorter than the first. Martin and Ron had no presentation to make, no powerpoint slides to show. Instead, they sat down with the wrapped bottles on the table in front of them, and explained to Ian Parker that they had brought with them ciders that were not on Tesco's shelves, nor were they on anyone else's. They described in mouthwatering detail the pressing of the juice and the creation of the finished cider, what it looked like, smelled like and tasted like.

Calmly Martin and Ron explained that Thatchers had a new range of products to offer, ciders aimed not at the low-cost market but at discerning drinkers, both men and women. As the conversation continued the buyer's eye returned repeatedly to the tissue-wrapped bottles, but only when the pitch was done did Martin pick one up and tear off the paper, to reveal the first bottle of Thatchers Katy, gleaming like an Oscar trophy.

Giving little away the buyer said he'd be in touch, and they drove off, wondering if they had done enough, and by the time they got back to Sandford he had convinced himself that they probably hadn't. Then they pulled in to the yard at Myrtle Farm and John came rushing out. The buyer had called, he said, almost immediately after the meeting, and put in an order. They wanted so many cases… In six weeks. Fantastic!

Now they just had to work out how they were going to bottle and label all that cider and deliver it on time, without having any bottles on site, or any labels, or even a design for the labels. A new chapter in the Thatchers story had begun.

Opening of Primary Process on 19th September 1997, with MEP Graham Watson.

THATCHERS
DRAUGHT CIDER

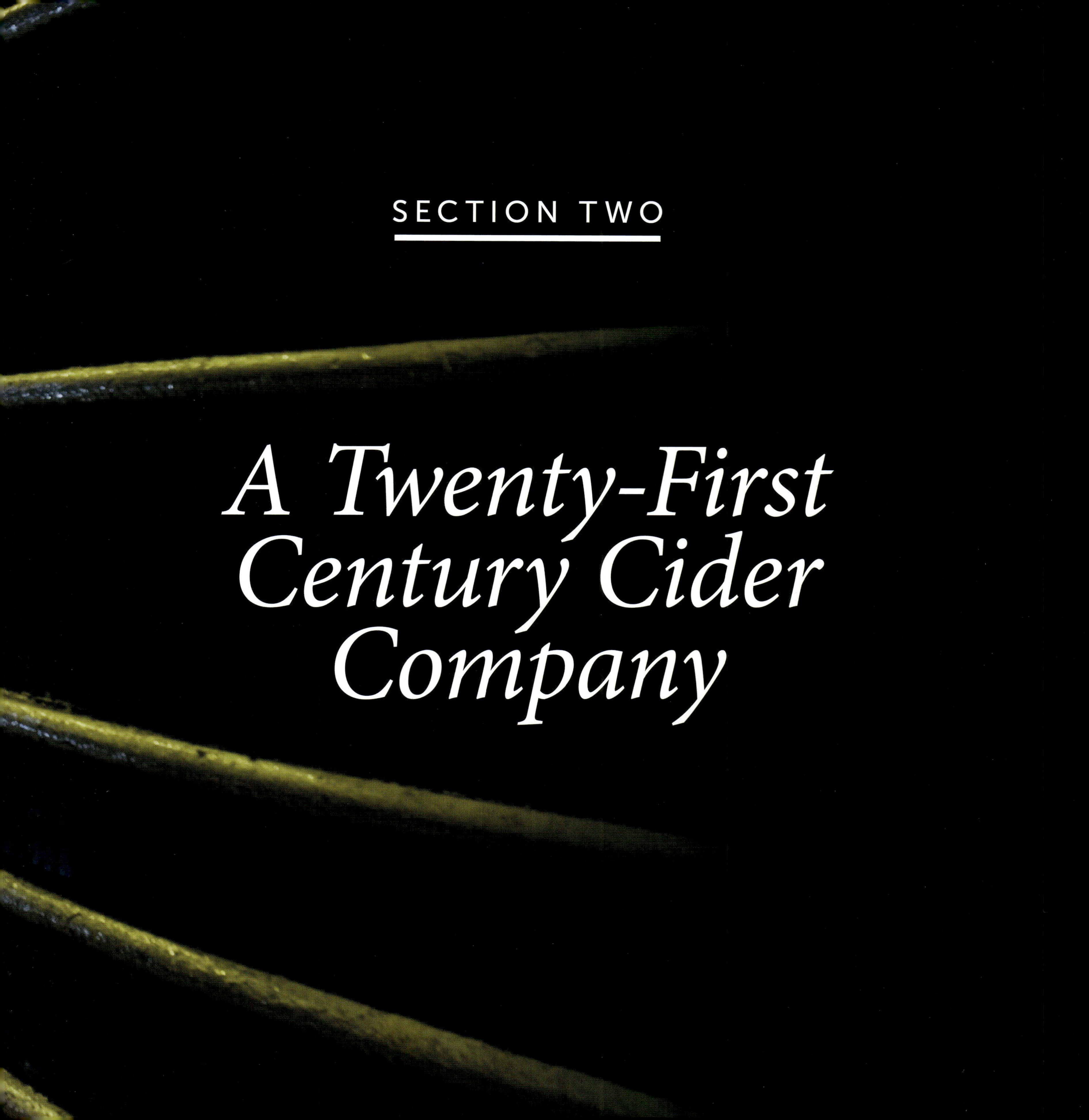

SECTION TWO

A Twenty-First Century Cider Company

Shiplate orchard in springtime.

Working solo (although with the vital support of Ann and trusted staff), John had proceeded cautiously over the previous ten years. Though not lacking in entrepreneurial spirit he still thought of himself as a farmer who made cider, and the farm was always there, ready to be put back to work should disaster strike the cider industry. He undoubtedly saw the potential for growth, but perhaps he knew too many former cidermakers to take anything for granted.

Martin was different. John sees in him more of his Uncle Reg, in an earlier generation the most successful businessman of the family, and no doubt those six years away at boarding school also made him think differently from his father. After the signing of the Tesco deal he was optimistic about the future, perceiving that Thatchers had something nobody else was able to offer and believing that further success would follow provided the growing team did everything right, down to the last detail. And if there are two things that separate successful businessmen from the rest of us they are self-belief… and attention to detail.

INVESTING FOR GROWTH

Having successfully sold cider to Tesco he knew that dramatic and sustained growth was possible, but this would require some serious investment and an acceptance that Myrtle Farm would from now on be devoted solely to the business of making cider. A second Bucher press had already been added, and in 1995 the company won a grant from the European Union to install a third press, along with a new apple mill and an evaporator to reduce the water content of juice so it could be stored as concentrate. A new bottling line was added too, and new temperature-controlled fermentation tanks which helped ensure the quality and consistency that supermarket customers demand.

To anyone who didn't know the cider industry well, the timing must have seemed odd, and not only because Chancellor Kenneth Clarke raised the duty on cider by 8p a pint in the budget that year. The appearance of the first alcopops in pub fridges around the country signalled the end of the boom in white cider and gimmicky bottled products, and for the next ten years mainstream cidermakers wrestled with falling profits and a lack of interest among the young. Suddenly the wisdom of Thatchers' move into a different market became clear, since their supermarket customers wouldn't touch an alcopop but were nevertheless interested in new, attractively-packaged drinks. Rather than stalling, sales of Thatchers' ciders continued to grow dramatically, with expansion into other supermarket chains including Waitrose. Production increased from three-quarters of a million gallons in 1994 to 1.75 million gallons two years later.

This growth was astonishing, but it rankled with John and Martin that their huge investment (£1.5 million including the grant) was for equipment that

The cider recipe is a century old, the bottling plant brand new.

for the most part was only used three months of the year, during the apple pressing season. Their first season of pressing with the new kit behind them, they began to wonder what else they could do with an oversized fruit press, other than let it stand idle. At the kitchen table they jotted some ideas down on an envelope, tore it in half and each went off to research the various possibilities.

APPLES ARE NOT THE ONLY FRUIT...

John's list included the laconic note 'blackcurrants'. A summer fruit, these were harvested long before the apples arrived, so Thatchers could press them without any inconvenience. Surely there ought to be some blackcurrants around that needed pressing, so John phoned an orchard owner who also grew blackcurrants and asked whether he knew of anyone who might fit the bill. The grower's reply was not at all what he expected. Everyone has heard of Ribena, the blackcurrant squash first created at Long Ashton Research Station and heavily relied upon during World War II as a source of Vitamin C. Well, it turned out that almost every commercially grown blackcurrant in the country was sent to the Royal Forest Factory in Coleford, Gloucestershire, where Ribena was (and is) made.

It so happened that Ribena's owners, Glaxo Smith Kline, were looking for someone new to press their fruit. Once again fortune seemed to be smiling. A note scribbled on an envelope and a single phone call had presented John and Martin with an unexpected opportunity. It was obvious that they should at least offer to press some of GSK's blackcurrants for them. They had the equipment after all, so why not? What they didn't realise was quite how many of the tiny dark berries were pressed for Ribena every year, nor indeed what kind of company they were dealing with. Thatchers may have been on the up, but it still only employed thirty five staff, whereas GSK was one of the world's biggest companies, with many thousands of employees in the UK and abroad.

So John and Martin arranged to see GSK's procurement manager James Wickham in Coleford, and duly explained that with their state-of-the-art fruit-processing plant they could press a hundred tons of blackcurrants a day. Now that sounds rather impressive, even among all the big numbers flying around, but the GSK delegation did not respond with the hoped-for enthusiasm. Sorry, was the answer, we need someone to press three hundred tons a day. Martin suggested that perhaps they could press a third of the crop and someone else the rest, but that wasn't going to work either. GSK had no intention of having a few blackcurrants pressed here and a few there. Either Thatchers agreed to press all of their fruit, at the required rate, or there was no deal.

Once again, father and son found themselves with a quandary. The GSK deal would make a fantastic contribution to the company finances, for a short, intense burst of work once a year. But their fabulous new press was not big enough. As they drove home across the Severn Bridge their first thought was of resignation. They would go back to the list and find some other fruit to press. But was there anything else out there quite like Ribena? Not even close. They drove along and, as always seemed to happen when they were in a car, John had an idea.

Perhaps, he suggested, they were looking at the problem the wrong way. It wasn't that they couldn't press three hundred tons of blackcurrants a day, it was simply that they were unable to do it with the equipment they had. All they needed to do was add two more presses, plus some other bits and pieces, and they would be all set.

A NEW PLANT – AGAIN

Given the huge investment they had just made in the company this was an outrageous suggestion, but Martin took it in his stride, and as they drove home they worked out what they would need, what it would cost – and how to pay for it. The plan was beautifully simple. If GSK were prepared to sign a long-term contract that would guarantee a certain income, they could afford to build a new plant to the multinational company's specifications. The implications for the cider business would prove immense.

Months of discussion followed, meeting after meeting where John and Martin would negotiate with fifty people from GSK. At Thatchers, decisions could be made in the car or at the kitchen table, but at GSK even the smallest decision had to be OK'd by so many people that it took weeks. To say this is not to criticise, but rather to highlight crucial differences between a small family business and a corporation, and to suggest why the family business model often works more efficiently. Farmers in particular have always needed to be flexible and opportunistic, adapting to the vagaries of weather and changing markets. They also need to be determined.

Eventually a deal was hammered out, the new plant was installed and in June 1997 the first delivery of blackcurrants arrived. It is quite hard to imagine three tons of blackcurrants, never mind three hundred tons, and as well as the sheer volume of fruit, the Thatchers team had to cope with being under the watchful eye of GSK's quality control inspectors.

By 1998 the transformation of Myrtle Farm was almost complete. The new production hall that covered much of the farmyard contained one of the most advanced fruit handling plants in the country, allowing the company, now Britain's fifth biggest cidermaker to press twelve thousand tonnes of fruit as opposed to seven thousand the previous season. Sales were up too, more than fifty per cent year on year; by June 1999 the company reported that it was producing around two million gallons of cider a year.

CIDER FROM ENGLISH APPLES

These impressive numbers beg a question. The late 1990s were by no means a boom time for the UK

Thatchers kegs are delivered to pubs up and down the country.

cider industry so how was Thatchers able to keep producing and selling more and more cider? Having the equipment to press vast numbers of apples was one thing, but selling all that cider was another matter. It required first of all cider that people wanted to drink, and secondly a sales and marketing strategy that succeeded in placing that cider in the hands – or pouring it into the glasses – of those people.

Around the West Country there was still a good market for traditional farmhouse cider, and Thatchers has never for one moment abandoned the bittersweet blends first created by William Thatcher – or, most probably, by his father Benjamin. It is telling of the way the company does business that you can walk into certain Bristol shops today and fill up a container from a barrel of Traditional, just as customers did in the 1970s.

But this was always going to be a fairly small market. Pubs needed something that offered their customers something slightly more refined, with a guarantee of long-term consistency, so a new kegging line was put to good use with a new draught cider. Filtered to a clear golden colour, Thatchers Dry retained some of the bite of traditional scrumpy while offering less tannin and a cleaner taste. Simple labels on bar taps caught the eye with 'Dry' – yes, you are thirsty and this cider will quench that thirst! Pubs took to the new kegs with alacrity, so that for drinkers in Bristol and the surrounding region Thatchers ciders soon became the most familiar on the bar.

The Thatchers name was also becoming among the most respected, it was to its great advantage that the company's apples were collected from English orchards, specifically orchards in the great apple-growing counties of Kent, Herefordshire and Somerset.

By investing in modern pressing equipment John's cidermaking team pressed apples and stored the juice as concentrate for fermentation later in the year, very different from the cost-cutting methods of their rivals. You don't have to be an expert to appreciate that cider made with vintage apple varieties, that have been checked for quality and pressed separately with the greatest care, will be of the highest quality. Once again, Thatchers' insistence on quality proved in the long term to make economic sense.

CIDER FOR WINE LOVERS

But it was the single variety ciders that really set Thatchers apart during the alcopop-crazed 1990s. While other cidermakers strived to make their products cheaper and sweeter, Martin and John led the company upmarket. To some extent the Tesco deal was an experiment for all concerned, because at the time neither the supermarket nor their newest supplier could have known what demand there would be for these single variety ciders in their shiny glass bottles – it wasn't as if anyone had gone out and asked consumers whether they fancied a glass of Katy.

With hindsight it is easy to see such decisions as

All manner of wildlife thrives in the orchards.

New buds open, a new season begins.

obvious, but at the time, in the day-to-day reality of running a business there is little to guide the decision-makers other than instinct. In the case of Katy, for example, the cidermakers were presented with a batch of delicious eating apples, and happened to have equipment that could get the best out of them; a supermarket buyer asked for something new and interesting and so here, almost from nowhere, was a wonderful new cider. While remaining a respected regional cidermaker, Thatchers now had a new identity, as purveyor of excellent single variety ciders – on a national scale.

We have already seen that cider has long been enjoyed by discerning Somerset people as a locally-produced alternative to wine, but on occasion producers have persuaded consumers far and wide to try it too. As long ago as the 17th century producers in the Three Counties had charmed royalty with Redstreak Cyder, which was reckoned to be as strong and flavourful as Canary wine, and which was so popular that an epic poem was written about it; three hundred years later a Somerset family business did something similar in a very different market with a sparkling perry called Babycham. In the austere 1950s this inexpensive alternative to champagne offered much-needed fun and glamour to a generation of women.

Although few products matched the popularity of Babycham there were nevertheless many examples of small companies achieving success through a combination of quality and canny marketing. Cider had often been promoted as an alternative to wine, or, as Whiteway of Devon once suggested in an advert, 'a veritable substitute for Hocks and other Still Wines', and now Martin set about branding Thatchers single variety ciders in this way. Beautiful glass bottles, colourful labels and elegant fonts all contributed to the image presented to shoppers, although none of this would have meant a great deal if the ciders inside weren't of the highest quality. And, as critics increasingly avowed, they were.

CRITICS GO MAD FOR THATCHERS!

In August 1998 a writer in London's Evening Standard picked out Thatchers as one of very few cidermakers offering an experience close to that of drinking wine. Among the brands covered in Andrew Jefford's review were Tremlett's Bitter, 'which combines clean cider-apple flavour with plenty of dryish tannic bite, making it a perfect mealtime red wine substitute', Somerset Redstreak, 'gentler, more elegant and white-wine-like', and Katy, 'with its scent of apple sauce and faint toffee-apple flavour'. Dabinett, meanwhile, was in 'a lighter, more quaffable style yet with a firm tannic undertow'.

Writing in The Sunday Times, Andrew Barr also picked out Tremletts Bitter and Somerset Redstreak, again comparing the ciders to wine. The latter, he noted, was 'lean and elegant, with a surprising length of flavour.' The author took this further, suggesting that Martin Thatcher had more in common with

A cidermaker's favourite sight: apples ripening in late summer sunshine.

colleagues in the wine industry than with his fellow cidermakers and pointing out that the new processing plant allowed apples to be pressed and fermented with minimal danger of spoilage. Controlled cool-fermentation and ageing in oak vats ensured quality and consistency, while also enhancing the flavour of the apples.

'We believe that cider should be soft and fruity, not hard and acidic,' Martin told him. 'It is part of making cider palatable.'

Once again a producer at the northern base of the Mendips was creating ciders that were 'strong, palatable and highly esteemed…' That same year Thatchers unveiled another bottled cider made from the Cox apple - a devil to press but producing a medium-sweet light cider that was considerably less strong in alcohol than Katy. Then came Spartan, a slightly stronger cider that was also made from the juice of a dessert apple.

Featured on the BBC's Food and Drink programme in May 2001, Spartan was given a firm thumbs-up by wine guru Oz Clarke. 'I think this is an absolutely perfect example of a modern cider,' he said during the show, 'high quality, clean as a whistle, refreshing, smashing stuff…'

Sold in half-litre bottles with bright painted labels, the range of single variety ciders now offered something for every palate, from light, fruity Cox's to the tannin-rich, powerful Tremlett's Bitter.

GOOD TIMES FOR CIDER INDUSTRY

In April 2004 Martin explained his philosophy to Bristol's Venue magazine, one that William John Thatcher would surely have shared. 'If you look at cider as a product,' he said, 'there's such a diverse range of flavours, character, audience appeal and alcohol level. More so than beer or wine. From traditional, cloudy, completely unfiltered cider to single varietal ciders that are presented in a very sophisticated way, there's a very wide audience. At times the cider industry has forgotten to tell customers what cider is…'

At other times, however, the cider industry has spoken out clearly and with pride about its products and, just as importantly, people have been prepared to listen. One of those periods had just begun when Martin made these comments. Having marketed cider as a glorified alcopop for more than a decade, the mainstream producers suddenly revealed to an astonished world that cider was made from apples, grown in orchards, and was in fact the most natural of drinks. A hot summer or two helped and with the odd gimmick thrown in – could you imagine Stan Thatcher serving his cider over ICE? – the mainstream cidermakers grabbed the attention of the nation, and as happens with these seismic shifts, everyone began drinking cider.

Having continued to invest in new plant since the turn of the millennium, Thatchers was particularly well placed to make the most of this surprise

WELCOME

THATCHERS
Handmade Pottery
3 handled mug
(½ pint) £11.00

CIDER
SHOP

THATCHERS
CHEDDAR VALLEY
THATCHERS
Gold

5
GALLON
Without
Barrel
5L
TRADITIONAL

THATCHERS
SINGLE VARIETY
PRINCE WILLIAM

The Thatchers shop continues to evolve, but the barrels haven't changed much.

development. Not only were the Bucher presses capable of pressing 20 tonnes of apples per hour, but juice could also be stored as concentrate for fermentation at a later date. With large volumes of juice and cider on site at any one time it was possible to meet new demands quickly, whether for increased quantities or for different blends.

In the autumn of 2004 Thatchers released its first perry, a slightly sparkling version of the venerable drink made with Conference pears. Martin was sanguine about the reasons for choosing this dessert pear over traditional perry pears, the most obvious being that the latter are not easy to come by. Importantly, Conference also fit the programme of working with fruit that were familiar to consumers, while also offering a more reliable product than perry pears, which can give unpredictable results.

STRIKING GOLD

Back in the mid-1990s John Thatcher was chatting with Mike Hughes of Bulmer's – at the time one of the most influential figures in the cider industry – when Mike said something that has stuck in his mind ever since. They were discussing the challenges facing the industry at the time, and he said, simply, 'We've got to make the cider easier to drink, John.'

This might sound a bit odd at first. What could be easier, we might well ask, than drinking cider? But John understood exactly what he meant. There are two kinds of strength in an alcoholic drink, the strength in alcohol and the strength in flavour and general feel. Some red wines are 13% plus but slide down a treat, while others are weaker but don't. We describe them as rough or heavy, which generally means that they contain a lot of acid or tannin, and the same can be true of cider. Proper Somerset cider like Thatchers Traditional or Vintage is strong both in alcohol and in terms of the drinking experience, mostly because of the high levels of tannin in bittersweet apples (about which more below).

And however much purists may protest that no other cider will do, it is a fact that any producer seeking to expand out of the local Somerset market needs to offer something different – of high quality, and with pride in its origins in the orchard, but less challenging. Single variety bottled ciders like Katy and Cox's offered one route forward, since they were far lower in tannins than, say, Tremlett's Bitter. Meanwhile the success of Thatchers Dry showed that there was scope for a draught cider that had lots of flavour and a softer mouth-feel than traditional scrumpy. Here was an area that deserved more research, and as the new millennium got underway the cidermaking team began working on a new blend. After exhaustive trials they came up with a blend that seemed right: fresh and appley, with none of the harshness that customers found such a challenge.

Something Thatchers had discovered over the previous decade was the effectiveness of keeping brands simple. The days of Mendip Magic and

With the onset of autumn the cider apples begin to fall from the trees whatever the weather.

Mark Townsend, one of Thatchers' longest-serving employees.

Sandford Superb were long gone. Katy, Dabinett and Somerset Redstreak were all named after apples. Thatcher's Traditional and Dry were… traditional and dry. So when it came to the new draught cider there was no need to think of some elaborate, rustic-but-modern-sounding name; what made this new cider distinctive was its pale, golden colour, and so it was called simply Gold.

True, this was a name that carried potential risks. If you're presented with something named after a precious metal you expect it to be not just good but excellent, and anything less than excellence would be punished by critics and consumers alike. Cider drinkers have finely developed taste buds – as other producers have learned to their cost – and it was essential that the new cider appeal both to new and established customers.

Launched in pubs in 2003, Gold was an immediate hit, and as the cider boom continued, Thatchers fought its way into the mainstream for the first time with this light, sparkling draught cider. Gold was unlike any previous Thatchers draught cider, lower in tannins than Dry but initially stronger in alcohol, and with a bright, effervescent quality. Clean, natural and flavoursome, Gold was exactly what the new generation of cider drinkers – both men and women – were looking for, and two years later saw this most palatable of ciders available both in no-nonsense two-litre plastic bottles and elegantly packaged in glass.

Martin's genius for marketing had found its best expression yet: the same palatable cider packaged in different ways for different markets. The classy glass bottle also made clear the connection between Gold and the more sophisticated single variety ciders, encouraging people who had just discovered cider for the first time to expand their horizons with something like Katy (which, incidentally, sold a million bottles that year). Highlighting the importance of Gold, the company spent £350,000 the following year on a new carbonator, to ensure that every glass of the cider – whether poured from a can or at the bar – was served with an attractive, sparkly ring of bubbles around the top. It wasn't just the cider that was subject to Thatchers' exacting standards, the bubbles were too.

THATCHERS ON TV

In 2010 the company began investing heavily to promote Gold, with billboard and TV adverts that built on the qualities we've been looking at in this book. The television ads are interesting because, like so much that Thatchers does, they managed to be simultaneously old and new, embedded in the countryside yet modern in style. And, as elsewhere, simplicity is the watchword.

As barn doors open to a soft West Country sunset, a voice in the first ad quietly asks viewers, 'What is it about Thatcher's Somerset cider?' We see a bee fly through the orchard to pollinate a flower, which dies back as its fruit begins to swell, then a hand picking an apple. 'Could it be the juicy apples that grow in the

Apple harvesting, 21st century style.

Meanwhile, Danielle Gaydorova uses the traditional approach...

Another kind of thatcher... Chris Barnes, of master thatchers Dunbar and Bunce, at Williams Barn.

orchards? Or is it that the Thatchers family has been cidermaking here for a hundred years.' The famous old vats come into shot, followed by a glass of bubbling Gold, and the unhurried voice continues, 'Who knows? But I tell you what, one taste of Thatchers Gold and you'll know it was made by people who care.'

The thought of how much each of these well-chosen words cost when broadcast on national television may have kept Martin awake for a night or two, but the impact of this simple, unpretentious advert was huge. There were no great claims and no half-truths; everything about the ad is absolutely authentic.

As with other aspects of the business, Thatchers' growth has meant drawing on expertise from beyond Sandford, with advertising agency Bray Leino coordinating print and TV advertising, and matching this with product labels and so on. What people nowadays call brand management. The 2013 TV advert they produced, also for Gold, was particularly innovative. Directed by Martyn Pick, whose work had been seen in London in the build-up to the Olympics, it took the form of a short animated film, in which a digital painting technique is used to create a soft, shimmering image. The 30-second film includes an appearance by William Thatcher, who might have enjoyed seeing himself thus portrayed in the most modern of media. A lot has happened since Thatchers Cider was advertised in writing on a barrel, resting on the milk stand outside Myrtle Farm, but it is the family that remains at the heart of everything.

Made by people who care about cider...

A RAT AND A GOBLIN

Back in the noughties, Gold and Katy continued to enjoy unprecedented growth, while new avenues were constantly being explored. Perry was relaunched as Pear Cider as the latter caught on in the wake of the cider-over-ice boom, and there were new ciders made by Thatchers but branded by breweries. The company had long seen collaboration as a positive thing, since the days in the 1980s when the family name was less than popular in some quarters and a wholesaler sold scrumpy from Sandford as Farmer's Tipple.

The first brewery-branded cider, Moles Black Rat, was named after the abiding legend that farmers would help along a slow fermentation by adding a rat to the barrel. Whatever the truth of the story, the cider proved popular (the fact that it was made by

A rise in duty awakens the Green Goblin...

Thatchers remained largely unknown) and further collaborative ventures followed. Dorset brewer Hall and Woodhouse entered the cider market in 2008 with its Badger Applewood Cider, but the most successful team effort proved to be Green Goblin, which was made originally for Wychwood Brewery, but which subsequently became part of the Thatchers range. Strong, oaky and with plenty of tannins, it was a cider from the old school and marvellously packaged too.

The goblin himself came to life in a startling series of adverts launched after a certain Chancellor imposed an untimely and highly unpopular 10% increase in the duty payable on cider. Promoting Thatchers' 'oakrageously oak-aged' cider, advertising agency Wild Colonial Boys created lively, satirical print ads, in one of which the green goblin cries, 'Had Alistair Darling tasted this, he would've slapped on a flavour tax as well!'

Coincidentally, Green Goblin outlasted both the Chancellor and his tax rise, which was scrapped shortly afterwards in the face of massive opposition.

KATY GOES ROSÉ

Ventures like this set the trend for the following years, with an evolving collection of specialised bottled ciders balanced by a smaller number of big-selling draught ciders. It would be impossible to do justice here to the full range of innovations and variations, but a few favourites do stand out. One such is Katy Rosé, a variant of the pioneering single variety cider which was launched in 2009 during a boom in the popularity of rosé wine. In a manner that will be familiar by now, this delightful summer drink was created through a mixture of chance and inspired thinking. There had been an issue since Katy first

Sunrise in the orchard: it will be a fine harvest.

started being pressed for cider, in that the red skin of the fruit made the finished result slightly pink. In the mid-1990s this was considered a fault to be remedied, and a solution was duly found, but a decade later, amid changing tastes, Martin and his team realised that the pink colour could actually be an asset.

It was one of those moments when you hit yourself on the forehead and cry 'of course!', but like so much else that Thatchers has done there was no shame in having not thought of this sooner – after all, nobody else had. Sold in 75cl bottles like wine but with lower alcohol, a delicious appley flavour and an affordable price tag, Katy Rosé was the hit of the year, a cider bought (primarily) by groups of women who shared a bottle, drinking from wine glasses.

Constantly innovating, yet remaining constant to the central values of a family cidermaker pressing English apples, Thatchers continued to grow, winning new contracts with pub owners like Wetherspoons and then investing in new plant, such as a kegging line capable of filling 450 kegs an hour, to meet demand. The success of the company's first TV advertising campaign saw production pass the four million gallon mark for the first time by the end of 2010, while turnover was over £22 million for the year – up from £5 million at the start of the decade. Meanwhile Thatchers' forays into foreign markets earned the company a Business Thinking award from HSBC – a fitting reward for years of bold and inspired brainwork.

CHANGE AT THE TOP

One of the essential characteristics of a successful family business is the smooth passing on of responsibilities from one generation to the next. As an example of the opposite case John Thatcher tells the story of a business owner who said to his son one evening, 'Right, there are the books, I won't be in tomorrow'. This may seem a bit childish but it is understandable. Every family business has something of the Shakespeare play about it, with different family members striving to take the company in their chosen direction. And who wants to leave their life's work in the hands of others?

John and Martin Thatcher have always been close although (as both are keen to point out) they by no means always agree. Martin grew up with John as his father and also as his boss, and as a youngster he probably looked forward to the day when he would have his own farm and give the orders himself. Instead he came back to work for his father, proving himself adept at sales and marketing and gradually reshaping the company into a modern producer of nationally recognised branded products.

Perhaps it was inevitable that he would steer the business into the 21st century, but that he was able to do so is thanks to his father's good sense and willingness to put the company's future before his own desire to carry on running things as he had for so many years. Subsequently neither Martin nor John could remember how or when the transition took place (in itself a sign

The apple harvest in full swing.

Thatchers staff, family and friends at the opening of the Jubilee Building.

of how smoothly it happened), but the new Millennium found the former installed as Managing Director with his father as Chairman. Ann, it should be noted, was also a valued member of the Board.

Subsequently John and Ann have both retired from the Board although, as we shall see, John's work in orcharding and pomology has continued nevertheless…

CELEBRATING THE PAST, BUILDING FOR THE FUTURE

In 2004 Thatchers celebrated the centenary of William Thatcher's move to Myrtle Farm and his first year of cidermaking there. It was a joyful occasion, not least because four generations of the family were present when John and local MP Brian Cotter tapped the first barrel of Thatchers Centenary Cider: Stan's widow Molly, John and Ann, Martin and his wife Anne, and their children, four year-old Eleanor and baby Peter. This celebration was as much about the future as about the past because, like a farm, a family business needs a strong sense of the future to thrive.

There must be a feeling that today's hard work will result in benefits a generation from now, or even two generations, because that way decisions can be made that favour long-term growth rather than short-term gain. The situation is quite different in a publicly owned company, where annual profits are all-important.

When John decided to make way for Martin as Managing Director, he was thinking about his grandchildren's inheritance, should they decide to become the fifth generation of Thatcher cidermakers – the sixth if we include Benjamin. The single principle underlying Martin's running of the company is that he wants to hand on a substantially better business to his children than the one he inherited from his parents. Believing that it is better in the long term for the company to stay in Sandford, where it has always been, than to move to a bigger or more convenient site, Martin has overseen the office staff's move to Sandford Farm (once William Thatcher's home).

THE JUBILEE BUILDING

Meanwhile he chose to invest in the future at Myrtle Farm, with a new building that he hopes people will enjoy working in – and looking at – for decades to come. In contrast to most structures of its kind, the Jubilee Building is a thing of beauty, with curving lines and floor to ceiling windows: a simple, elegant edifice designed to be both sustainable and flexible.

Named in honour of the Queen's Diamond Jubilee, its opening marked by a visit by the Earl and Countess of Wessex in 2014, this spacious production hall incorporates many of the features we would expect from a progressive 21st century business, from energy-saving technology to rainwater harvesting. Heating is provided by a biomass boiler on site, for which there is no shortage of fuel. With many thousands of apple trees pruned every year, there is a plentiful supply of wood, which is chipped and then fed into the boiler.

The rich palette of colours in a Thatchers orchard.

Peter Golding shakes the trees.

There are worse places to be on a sunny morning... Chris Kennell collecting apples.

Across the wide floor of the Jubilee Building the future of Thatchers will be mapped out in ever-more-efficient packaging systems, which will be able to bottle, keg and can as much cider as the company's customers want – wherever in the world they may be.

In the same spirit of planning for posterity, Martin willingly commits the business to long-term contracts with orchard owners. He is looking not just at this year's supply of apples, or next year's, but twenty or thirty years into the future when Peter and Eleanor may be running the company.

Although cut from rather different cloth than Stan, Martin nevertheless believes strongly in the culture of Thatchers, a culture which transcends pure economics. An accountant could no doubt strip costs out of the business, making it more efficient and potentially more profitable in the short term, but while Martin is certainly keen to increase efficiency through innovation, he understands that there are costs that must be accepted – and he accepts them gladly – if the company is to prosper long-term.

Part of the value of the company may be in its brand equity and part in the quality of the product, but there remains a third factor which is almost impossible to quantify: the good will of staff, customers, neighbours, suppliers and all the other people involved with the business. To Stan Thatcher the making and sharing of cider was all about friendship and conviviality, and, just as local customers still visit the shop to fill a flagon from the barrel, this spirit still infuses the company.

So it's worth having the cidermaking team take time off to show local people around on Open Day, and it's worth investing a huge amount of time and effort into the redevelopment of the Railway Inn, to provide a social hub for the village. An accountant might raise their eyebrows at the costs involved – unless he or she is an accountant with a taste for cider.

A WINNING TEAM

The same is true in the way both John and Martin have managed the company's staff. Things have changed since the days when Ann handed out wage packets through the kitchen window on a Saturday lunchtime, but the underlying principles remain the same. Thatchers may employ over a hundred people nowadays, but each member of staff is valued highly, and given as much opportunity as possible to grow as the company grows.

Take Neil Day, who first arrived as a thirteen-year-old schoolboy with a Saturday job washing lorries, before leaving school at sixteen and coming to work for Thatchers full time. Having left briefly to train as a quantity surveyor he returned and proved himself invaluable, overseeing cider production and eventually working alongside Martin as Operations Director. The boy who left school with five GCSEs found himself with seventy-odd people working for him… including several PhDs.

John is a firm believer in finding the right people

when they're young and has been known to invent a job where none existed for a lad who looked like he was made of the right stuff. But Thatchers has also proved adept at attracting high flyers with long careers in the industry already behind them, people like Technical Manager Paul Dockerty, who is arguably the best cider scientist in the country.

'You've got to be an opportunist in this world,' John says. 'If someone comes along who's good you grab them.'

You have to keep them too. When Gary Delafield first started work at Thatchers he was on the production line, packaging cider. That was in the 1990s, and bottles were filled by hand from the barrel. Then, as we have seen, the company expanded massively, and Gary worked his way up to become Production Manager. Years later he remembered the moment when Martin came to him on the production line and invited him to come and join the Friday tasting that week, on the basis that Gary would benefit from an insider's knowledge of the product he was bottling. Was this just a thoughtful managerial gesture, or had Martin spotted Gary's potential for a future in cider production?

MICROMANAGEMENT? NO THANKS!

Either way, both John and Martin have often trusted their instinct when it comes to the people who work for them, and who they work with, and have a day-to-day management approach that involves passing

Prize-winner: Neil Day, Operations Manager.

responsibility as far down the line as possible. John tells the story – almost shuddering as he does so – of a farmer friend who is so controlling that if one of his tractors gets a puncture, he has to decide how, when and by whom it is to be fixed. In John's case, the first he'd know about it is when the bill for the repair arrives. Even when you're talking about new equipment worth millions it is the relevant manager who works out exactly what is needed, reporting back to the MD with a plan, a product and a price.

Martin has gone out of his way to study business culture, both through his dealings with companies like GSK, and more systematically, when he travelled the world on a Nuffield scholarship and visited more than

Tony Franklin at work.

a hundred companies in Australia and New Zealand, China and the United States, as well as Europe. He believes there are two approaches to running a company like Thatchers. You either have loads of rules and systems or you say to people, look, this is where we're going, now let's get on with it.

While the former method might give the senior management greater control, it is bound to stifle creativity, leaving staff with no incentive to do anything. The flexible approach, meanwhile, leaves people able to make decisions and take the initiative. The key is to nurture or bring in the right people, people of a similar ilk, and allow them as much freedom as possible. Yes, mistakes will be made, but the benefits of this approach are clear.

In an age of zero hour contracts and constantly changing staff, Thatchers has built a team of highly experienced and motivated people who are capable of amazing feats. It was extraordinary to discover, for example, that the production team had installed the kegging line in the Jubilee Building themselves; between them they know exactly what every bit of hi-tech equipment does, they know how it all fits together and, generally speaking, they know how to fix it. But perhaps we shouldn't be surprised. This is how things always were on the farm, when John, Stan or William expected staff to sort out problems without calling in expensive help from outside. Today, everyone in cider production has the technical know-how to look after the machinery, and is encouraged to suggest improvements or efficiencies.

The company has also excelled over the years in maintaining relationships with customers, suppliers, like-minded people in other areas of business, and friends in the media. After Oz Clarke enthused about Thatchers on BBC television Martin approached him to lead a Tutored Cider Tasting at the BBC Good Food Show, in November 2004. In a fitting end to a year of celebrations, 10,000 people visited the Thatchers stand during the five-day show at the NEC in Birmingham, with several hundred of them attending a tasting session hosted by Oz and Martin.

For Martin, though, everything comes back to Myrtle Farm and the people who work there. Do they wear their company polo shirts or jackets at the weekend? If yes, then they're proud of where they work, and why are they proud? Not because of something Martin or the management have achieved but because they, as individuals and as a team, have created products that people know and love.

INSIDE A MODERN CIDER COMPANY

One of the problems of writing any kind of history is knowing when to stop, and with a fast-growing business like Thatchers the issue becomes particularly tricky. In less than twenty years the company has gone from small-scale traditional cidermaker to a major force in the UK and international industry. Where apples arrived in sacks and the finished cider was bottled (if it was bottled) by hand, we now see one of

The finest cider straight from the farm.

ON
5L
Without Barrel
£34.75
£8.55
Without Container
£7.65
2.5L
£4.85
Without Container
£3.83
2L
£5.56
Without Container
£3.06
TRADITIONAL
THATCHERS
£8.50
THATCHERS Gold

A lot has changed at Myrtle Farm – but this hasn't.

the most advanced cider works in the country.

So what happens to a batch of apples today when it arrives at Sandford? First, the fruit is unloaded into one of four pits – increased a few years ago from two so that different apples (and pears) can be kept segregated. From there apples are carried by water to be milled and then pumped into one of five Bucher presses, the smaller one that was installed all those years ago, and the four larger machines that have been added to meet demand since.

From the moment it enters the works, the fruit is protected from any chance of spoiling. It is carefully monitored for quality and crushed with just the right amount of pressure. The juice is piped immediately to one of twenty-four temperature-controlled fermentation tanks, where the process of fermentation is tracked to the minute by a computer system that displays every relevant piece of data. When the yeast has done its work the cider is immediately filtered, removing any remaining active ingredient and ensuring that no further fermentation or other change in the cider takes place. From here different ciders take different routes, but those made with traditional bittersweet blends go straight into the giant oak vats that remain the most compelling feature of the works, to mature for a month.

THE BEST OF THE NEW...

We'll come back to these vats, but for now let's follow the finished cider, because it's in the packaging and distribution of cider that Thatchers has perhaps made the greatest advances in recent years. Here as elsewhere at Myrtle Farm the aim is not to make things bigger or more complicated, but wherever possible to simplify processes and to ensure that every drop of cider is enjoyed by whoever ends up drinking it, wherever they are. Once upon a time this meant pouring the cider into a clean container at the shop and warning customers to drink it fairly quickly, before any oxidisation could occur. Nowadays, by contrast, the customer might buy their cider in a shop in Sydney, Australia, weeks after it has left Sandford, but they still expect – and quite rightly – to derive the same pleasure from their favourite tipple as a customer buying from the Thatchers shop.

Packaging, therefore, is crucial, and it will come as no surprise to learn that Martin has spared no expense in acquiring the best equipment available for the job. The bottling line is from Krones, the German company that leads the world in packaging innovation, and it comes with attention to detail built in. To take one example, when you fill a bottle with cider there's a point where the liquid comes into contact with air, which may eventually spoil the taste if the bottled cider is kept for a long time. But at Thatchers the air is expelled from each bottle before it is filled, so that the cider leaves Sandford having never, at any point since the apples entered the building, been exposed to the air.

Then there are automatic cameras to ensure that

Blue tits and other birds keep pests under control – they're welcome!

each bottle is lined up correctly for labelling, and even a laser to mark the code on each individual bottle; a printed code can be erased or altered, but a code etched by laser makes it possible to identify each and every bottle with absolute certainty. As an aside it is worth mentioning that at Thatchers the labels are stored in climate-controlled conditions, so that each label is in the best possible condition when it is applied to a bottle. You may never notice a perfectly-placed label, but a crooked or damaged one will make an immediate and less than positive impression.

The kegging line is even more impressive since it features something William Thatcher could never have imagined gracing his farmyard: a pair of robots. This isn't quite 'Iron Man', in fact the twin machines have a limited but important function, lifting empty kegs onto the line and then stacking the filled kegs onto pallets at the end. This saves the forklift drivers a great deal of repetitive work and reduces the risk of damage or accident.

AND THE BEST OF THE OLD

We've come a long way since that first advert in the Bristol newspaper, but here's a curious thing: although the presses are Swiss and the bottling line German, and even though so much of the modern cidermaking process is computer-controlled, the company's bittersweet ciders are still matured in great oak vats that probably pre-date William Thatcher's move to Myrtle Farm. It's hard to say exactly when they were built, because each vat had several 'lives' before it reached Cheddar Valley and then moved up the road to Thatchers in the 1980s, but they are almost certainly Victorian.

Where they were built, and by whom, is a mystery, but to work with oak on that kind of scale takes tremendous skill. Were they constructed by coopers, as smaller barrels are? Or were they perhaps made by traditional ship-builders, who found themselves in the later 19th century losing business to iron-hulled ships? Thatchers staff enjoy telling visitors that the oak in the vats only narrowly avoided being turned into the warships of Nelson's navy – those 'walls of oak' that protected Britain from Napoleon – and although this is said in fun there's a serious point underlying it. Whoever built the vats used wood from oak trees of a stature that would be difficult, if not impossible, to find today; the trees were carefully selected, felled and cut into planks, each one up to ten metres long. Working by eye and with tools that must have been primitive by today's standards, the builders shaped the hard wood, assembled the vats and bound them in iron, intending them to last.

And lasted they have. The vats were the best available then, and they remain the best today. Since they were reassembled at Myrtle Farm none of them has been empty: they are filled with cider, which is usually matured for a month, then emptied and immediately refilled. This way the wood doesn't dry out, which might cause it to split or crack. You can rest assured that at any time, day or night, summer or winter, the vats at Myrtle Farm are full of delicious bittersweet cider, and that every Friday, whatever the season and whatever the weather – in difficult times as in good – the cidermaking team will be up on top of the vats, testing the cider.

THE FINE ART OF TASTING CIDER

This is a tradition that has gone on as long as the vats have been here (and which no doubt happened elsewhere before that). You might think that in the 21st century testing the cider involves analysing it with scientific equipment, but there is only one way to properly test cider, and that is to taste it.

You reach the top of the vats via a walkway from the neighbouring room and first time around it is a disconcerting experience. Of course there are guard rails so that no one is in any danger of falling off, but it is still a long way down to the floor; more reassuring is the thought that you are standing on top of 77,000 litres of cider, made from the juice of fine Somerset bittersweet apples – Dabinett, Tremlett's Bitter, Somerset Redstreak and the rest. Strangely there are no shortage of willing tasters, but it is a responsible job so the most experienced members of the team are there – Gary, Paul, Quality Manager Richard Johnson and one or two others – each holding a long-stemmed glass.

When everyone is ready, a trapdoor is opened in the top of the barrel and a jug is lowered into the gleaming liquid that fills the vat right to the brim; you have to get that jug properly down into the cider, so whoever is holding it is expected to come up with an arm wet to the elbow with cider.

A good glassful of cider is poured all round and everyone stops chatting and focuses keenly on their glass. As with a wine-tasting you put your nose to work first, then take a good look at the cider and give it a swirl around the glass. Finally, you take a drink, holding it in your mouth for considerably longer than you would normally, before swallowing it.

This is only one mouthful of cider out of thousands of litres, but this single drink tells the experienced tasting team exactly what they need to know about the whole batch. So what are they looking for? Essentially, they need to establish that the cider is fully mature and ready to drink, but how can they tell? This is difficult to explain. They're gauging not only the flavour of the cider but the full experience of drinking it, from the moment it touches their lips until a good few minutes after swallowing. They're judging sweetness, bitterness, acidity. They're asking, what does the cider feel like? You can't translate a lot of this into words, but there's a sensation of warmth in a good, mature cider. There's a feeling, as John Thatcher likes to put it, that the cider has 'knitted together'.

If the contents of a vat is ready to go, it goes. If not, it stays until the following week. The team move from vat to vat, allowing several minutes between each sample and filling the time with informal discussions about upcoming orders, deadlines for this and that,

There's only one way to test cider... Master cider maker Richard Johnson hard at work.

supplies needed and so on. Martin only misses the Friday tasting if he is away on business, so it's a good opportunity for him to catch up with the cidermakers, swap news and bounce ideas around. There's a good feeling on top of the vats. Here, you feel, is a group of people who are where they want to be, making some of the best cider in the world.

The tasting team were understandably chuffed when, in 2013, Thatchers Vintage was awarded Champion Cider at the bi-annual International Brewing Awards; the cider the judges enjoyed so much had matured in one of the great oak vats and had been passed ready for drinking by the discerning cidermakers.

THAT FRIDAY FEELING

To leave the works, the visitor simply unlatches a wooden gate, but just before departing there's one more shed to have a look at. It's a smaller room than most, with stainless steel storage tanks labelled Dry, Medium and Sweet in the old way. The cider in here has been fermented like the rest but has bypassed the filtration system. It is cloudy and dry in flavour, just as Stan Thatcher made it and just as the regular customers at the shop still like it.

The barrels lining one wall of the shop are more pristine than in the past, but the set-up is in principle the same as in Stan's time. Around five o'clock on Friday people start to arrive, a man in ripe middle age swinging a five-gallon barrel by the handle, then two young women. Another car pulls in, and another, and the customers walk purposefully into the shop, to emerge a few minutes later weighed down with the delicious tawny cider that has given so many people so much pleasure over the years, and smiling at the prospect of that first sip.

Much has changed at Myrtle Farm but the old hayloft is still there, above the shop, while the dovecote across the way is used for storage. John and Ann live in the farmhouse, just as the farmer always has, with a sunny conservatory the only major addition. An air of calm reigns in the house, and you wouldn't know that a major cider works is in full-scale production less than a hundred metres away. And while John and Ann live at one corner of the site, Martin, Anne and the children live at another corner, just beyond a stack of empty kegs that are waiting to be filled. Martin may not have become a farmer, but he has a farmer's care for his business; he won't miss a Friday tasting if he can possibly help it, and should the duty manager need a hand on Sunday morning he'll be there, getting stuck in.

Her Royal Highness The Countess of Wessex at the opening of the Jubilee Building.

SECTION THREE

Apples & Orchards

Amazing what you can do with apples...

Specific apple varieties were being discovered and propagated long before the Romans came to Britain and planted the first orchards of dessert apples to grow on these shores. You can – as Thatchers has ably demonstrated – make cider out of eaters (or cookers, for that matter, or even crab apples) but the great cidermaking cultures of Northern Spain, Brittany, Normandy and the West Country owe their richness and longevity to a different kind of fruit. The cider apple was probably first developed in Moorish Spain and delivered to us via Basque and Norman traders, and from these distant beginnings thousands of varieties have emerged, flourished and, in many cases, vanished again.

Because apples do not grow true from seed they have to be propagated by grafting, and over time varieties tend to lose their vitality. At the same time the random nature of apple reproduction means that new varieties pop up unexpectedly (as the Granny Smith apple did, in an Australian compost heap).

For centuries Somerset cidermakers used a vast range of different apples, some excellent like the Court of Wick Pippin, others less so, until in 1903 the National Fruit and Cider Institute (later the Long Ashton Research Station) was set up just outside Bristol, to study cider apple varieties scientifically. The following year staff pressed juice for cider for the first time – the same year that William Thatcher made his first batch – and by 1910 the legendary cider scientist BTP Barker and his team had analysed the juice of more than 2000 apple varieties. For the first time cider apples were defined according to the now familiar classes of sharp, sweet, bittersharp and bittersweet.

The research confirmed what the best cidermakers had always known: the best cider is made with vintage apples, those with the ideal balance of sugar, acid, and tannins. More sugar means a higher level of alcohol, which in traditional cidermaking also helps the cider keep better. All apples contain malic and lactic acid, and their balance in a cider apple contribute both to the flavour and the health of the cider. The tannins that give traditional Somerset cider its rounded flavour, mouth-feel and amber colour are naturally occurring phenolic compounds that protect apples from disease and injury. If the skin of an apple is broken, these complex molecules oxidise and turn brown, covering the exposed flesh in a layer of thick, brown juice that protects the fruit from attack by fungus or insect. When the apples are pressed for cider, the same molecules add body and flavour.

Somerset cider is so distinctive in part because the apples used to make it are themselves products of the Somerset landscape and climate. This doesn't just mean that each apple ripens in the gentle, well-watered West Country, but that the varieties themselves were born of the region's soil. They appeared here first, and they grow here best, and different cidermakers blend their favourite varieties, each in their own way, to produce ciders that reflect

the region, the county, the village, and even the individual farm or works where the cider is made.

DABINETT

Of the apples employed by William Thatcher and now blended to create such memorable ciders as Thatchers Vintage, perhaps the best all-rounder is Dabinett, a full bittersweet which ripens late in the season and can generally be relied upon to produce a good crop (not something you can say about every variety). Equally successful as a standard tree and in a bush orchard, Dabinett was first discovered growing in a hedge in East Lambrook, Somerset, by Mr William Dabinett, and was propagated with great success by a local nurseryman. This is by no means an unusual story, since the 'gribble', or self-sown seedling, occupies an important place in the culture and history of Somerset cidermaking.

Until the Long Ashton Research Station was established, farmers would often spread their pomace in a gribble field, allowing livestock to eat it, then leaving fencing off the area so that the seeds could take root. Most of the resulting seedlings were useless, but out of a thousand or ten thousand seedlings one might prove to be a new and fabulous cider apple. This the cidermaker would nurture and then propagate by grafting, which is exactly what happened with Dabinett. According to expert pomologist Liz Copas, the first Dabinett was probably a seedling of Chisel Jersey, and it is a classic cider apple, medium-sized but weighty, conical like all the Somerset 'jersey' apples and giving a sweet juice with a soft, full astringency. The cider it produces is usually excellent – flavoursome and full-bodied, which makes it ideal either as a base for a more complex blend, or as a single variety cider.

SOMERSET REDSTREAK

Another variety that Thatchers has grown in huge quantities over the years is Somerset Redstreak, which Liz Copas suggests is an old variety from south east Somerset. It is an apple that really came to prominence with the first bush orchards, as it fares better in bush form than as a standard tree, and it crops well every other year. Unlike Dabinett, the bright red fruit is ready early in the season and usually needs to be pressed around the end of September, and in a good year can produce a pleasant Single Variety cider with soft tannins. Anyone who wonders why single varieties tend to come and go from the Thatchers list should bear in mind that a particular variety, orchard or even tree can produce fruit of wildly different character from year to year, depending on the weather. One year a batch of cider made with a certain apple might be so good that it simply has to go out as a single variety, while the next year it may be better suited to blending. Such are the variables that the cidermaker has to deal with, and which make consistency such a challenge.

Planting for the future bears fruit for generations to come.

TREMLETT'S BITTER & HARRY MASTERS' JERSEY

The same is true of Tremlett's Bitter, another Thatchers favourite and perhaps the most sophisticated of the bittersweets. Its tannins are hard, the flavour bitter; unless you're acclimatised, it's going to be a challenge to drink as a single variety, but for the true connoisseur Tremlett's Bitter is unbeatable. Since it first appeared in the Exe valley in Devon, this variety has been popular in neighbouring Somerset, the bright red conical fruit with white woolly flesh ripening early to give us some of the most characterful cider around.

Nobody knows quite why the 'jersey' apples have that name, but have it they do, and Thatchers' vintage ciders generally include the juice of at least one: Harry Masters' Jersey, which is named after the man who raised it in the Somerset village of Yarlington Mill in the mid-19th century. According to legend Harry owned the mill itself, but he may have been simply the local nurseryman. Either way, his name lives on alongside his apple, adding flavour spiced with Somerset lore to countless bottles of cider.

COX'S AND KATY

Producers in the eastern counties of England have traditionally used dessert and cooking apples in their ciders, principally because those kinds of apples were being grown anyway to supply markets in London. These ciders have much lower tannins than their counterparts in the West Country, but can be fiercely strong in alcohol – more like apple wine than cider – on account of the high level of sugar in eating apples.

Cox's Orange Pippin was first raised in about 1825 by Richard Cox, a retired brewer from Slough, and it has been cherished ever since for its deliciously sweet, intensely aromatic flavour and soft creamy flesh. The best-known English apple, Cox is a staple in eastern counties ciders, and has produced for Thatchers a delicious single variety.

Better still is Katy. Originally named Katja, this red-skinned, pale-fleshed eating apple, a cross between James Grieve and Worcester Pearmain, had been propagated in Sweden in the 1940s. Growers in Kent planted the variety in the 1980s at the request of supermarkets looking for attractive new fruit, but a change of mind left growers with a harvest nobody wanted. So there was a measure of opportunism in Thatchers' decision to buy up a load of fruit and press it, but the results were phenomenal, and today Katy apples are grown alongside the Somerset cider varieties in Thatchers' own orchards.

THE FIRST BUSH ORCHARDS

Research at Long Ashton transformed cidermaking in Somerset, enabling producers to plant specific varieties that could be blended together to create the best possible ciders. This was a golden age for the 'show cider', such as the champagne cider William Thatcher liked to make, vintage bottled cider made

One good shake and the apples come down.

Planting a new orchard at Thatchers.

in small quantities and with the utmost care. But the research did not stop with the identification and analysis of apples. Some of the research station's greatest achievements were in orcharding, both in promoting the best varieties for standard orchards and then, after World War II, in the development of bush orchards.

Cider apples are grown by grafting twigs of the desired variety onto a root stock of another variety chosen for its strength and vigour. Traditionally these were all standard trees, which were planted widely spaced so that livestock could graze around them. But the trees took fifteen years or so to reach maturity, and the yield of even the heaviest croppers was low when you consider the amount of land needed. So after World War II LARS began researching a new type of orchard, with rows of dwarf trees instead of standards, and after years of work were able to offer cidermakers varieties that retained the best qualities of their ancestors, but were ready to harvest after only three or four years. Although bush orchards required much more looking after than standards, yields were double or even triple.

As Long Ashton researchers and local cidermakers like Stan Thatcher were equally aware, there was another urgent reason for planting new orchards. Many of the old standard orchards, planted when farmers were still making cider for their workers, were past their best, while others were being grubbed up to make room for grazing land or housing. In short, the supply of apples was drying up, which offered an opportunity to plant a new generation of more easily-managed, highly productive bush orchards.

When Martin was seven or so he helped his father plant Thatchers' first bush trees at Webbers Orchard close to Myrtle Farm. In those days tree-planting was a laborious business, since you had to mark out the rows with string, and the position of each tree with a bamboo cane. To plant a tree you dug a hole with a spade that had a heavy pointed blade, set the sapling in place by hand and then filled in the hole, so if you planted a row of a hundred trees in one day you were going well. Yet compared to the old standard trees, the rapid growth and high yields of a bush orchard were almost miraculous. By the time John was ready to increase production after Stan's death, a good proportion of the apples were coming from the company's own orchards.

PLANTING CONTINUES...

Planting continued as demand grew, so that by the late 1980s Myrtle Farm was surrounded by flourishing apple trees. By the end of the next decade there were about 180 acres of orchards containing many thousands of trees, from Webbers to the more recent Nye Road, Shipham and Sacofs Orchards (all planted mid-1990s).

From time to time particular orchards have been commemorated in ciders that have born their name, such as the company's first single orchard cider,

Christon Orchard, which was launched in 2005 as a companion to Katy and the newly-released Vintage. These ciders were sold in sleek green 75cl bottles, similar in size to wine bottles but distinctively streamlined, reflecting in their design Thatchers' desire to emulate the high standards of wine producers while remaining determinedly different.

The orchard itself is a familiar sight to travellers on the M5, lying just across the motorway from the Webbington Hotel and Crook Peak; first planted with standard trees in the 1920s it was then replanted by Thatchers near the end of the century, with Somerset Redstreak, Katy, Morgan Sweet and Ashton Bitter growing alongside older varieties like Porter's Perfection and Harry Masters Jersey. For the single orchard cider each variety was pressed and fermented separately, then the ciders blended for the desired flavour.

JOHN THATCHER SAVES BRITAIN'S ORCHARDS

Apple growers everywhere were baffled and alarmed when, in March 2004, the national government announced details of how it planned to administer the new EU Single Payment Scheme, which subsidised farms by area rather than output. A vast survey was announced, a modern equivalent of Domesday, in which every hectare of agricultural land was to be mapped; subsidies would be based on this map, not only that year but in perpetuity. And orchards, even those containing only one old tree, were not eligible.

In other European countries orchards were included in the scheme, just not in Britain, where growers faced a stark choice: either accept that their land would be ineligible for farm subsidies, *even if there was a future change of use*, or grub up their trees – bizarrely, vacant land was eligible, so long as there wasn't a single fruit tree growing on it.

With wry humour, The Sunday Telegraph noted that the West Country was likely to see 'the biggest bonfires since foot and mouth, only they will smell better.'

Meanwhile John had gone to work, on behalf of the NACM and fruit growers everywhere, trying to persuade the government to change its policy, but Westminster showed no interest. For a time all appeared to be lost, until he went along to the annual meeting of the European Cider and Fruit Wine Association (AICV) that summer. As usual an official EU representative was there, and when John explained to him what was happening with British orchards, the man from Brussels promised to see what he could do. The upshot was that the wording of the policy was altered, and a great number of orchards – particularly old traditional orchards – were saved.

Further orchard plantings followed in the vicinity of Myrtle Farm, with 70,000 new trees established on one 180 acre site at Shiplate, not far from the existing Christon Orchard. Back in the 1970s it would have been almost unimaginable to plant trees on this sort

Few people have done as much for orcharding as John.

Like all cidermakers, Martin is happy to see a bumper crop.

of scale, but in the intervening period great leaps in progress had taken place. The first development saw a post hole digger mounted on the back of a tractor, with a burly chap pushing it down into the ground from above. This improved efficiency no end, but then came a second improvement with the digger moved to the front of a JCB and mechanical power used to make the hole. By the time the orchard was planted at Shiplate it was possible to plant 10,000 trees a day, with digging machinery following a GPS signal to get the rows even; Peter and Eleanor helped out at the weekends, placing guards around the trees.

APPLES: THE NEXT GENERATION

As a company Thatchers had been a member of the National Association of Cidermakers (NACM) for some time, and in 2001 John was appointed Chairman of an industry body that includes all the major cider producers. He used his chairmanship to press for increased investment of time, energy and money in an area of cidermaking to which he was himself devoting more time and energy: the cultivation of cider apples.

In the early 1990s he and Ann had travelled to Australia for the first time, and subsequently became regular globe-trotters, with John taking any and every opportunity to observe developments in fruit growing. In country after country he saw new methods, new machinery and new varieties; he saw increased efficiency and higher yields. Meanwhile a succession of British governments had gradually cut research funding for fruit and cider at LARS, until that remarkable institution finally closed its doors in 2003, a hundred years after it opened.

The writing had been on the wall for a decade at least, and some years earlier John Thatcher had planted saplings from LARS in an Exhibition Orchard of some four hundred varieties. The loss of the research station was a major blow, he knew, as there was no other scientific institution in the country that specialised in cider, and hence no one to continue the work that had changed the cider industry so much, and for the better. Who would now keep trying to increase yields, improve disease resistance, and continue the endless quest for new varieties? Where would the next generation of orchards come from?

Having been around cider for so long, John appreciates better than most how essential it is for the industry's future that new varieties are found and propagated. So he took it upon himself to continue research begun at LARS, working alongside NACM orchard adviser Liz Copas. Few people know more about Somerset cider apples than Liz, who worked for many years at Long Ashton and compiled the first Somerset Pomona.

Research into the next generation of apples began in the 1980s, mostly because unforeseen problems were occurring with the new bush orchards. Some varieties were becoming biennial, or cropping unreliably, and yields were not always as good as had been hoped. The worst problem was that the early and late-maturing varieties were the least successful, so that some cidermakers found that all their apples were arriving at the same time, in October.

So Liz and her team began a research programme – a quest, you might call it – to find and propagate new early maturing varieties. They were looking for varieties that cropped precociously and reliably, which were relatively disease-resistant, and, most important of all, produced juice that made good cider. Oh, and they needed to be easy to press too.

With this wish-list they set about making crosses between the most reliable cider apples (Michelin, Dabinett) and early-maturing dessert apples (James Grieve, Worcester Pearmain). The first seeds were sown in 1984/5, and four years later twelve hundred seedlings were selected and propagated. A decade passed, then the most-promising forty-six new varieties were planted in commercial orchards around the country, including Thatchers. By the time the final selection of the best twenty-nine varieties was made in 2007, Long Ashton Research Station had closed and houses were being built on its former orchards.

AN APPLE FOR A PRINCE... AND AN ORCHARD OF VINES

John Thatcher persevered with the experiment, planting rows of new varieties – early maturing bittersharps like Angela and bittersweets such as Lizzie. The most celebrated of the new apples was undoubtedly Prince William, a Dabinett cross. As a young man the Duke of Cambridge told the world that cider was his favourite drink, so Liz had the idea of naming the apple after him, and in 2008 Thatchers produced the first bottles of soft, subtle Prince William cider.

Prince William has proved a popular single variety cider, demonstrating that it was worth investing time, money and effort over twenty years or more and, despite the lack of publicly funded research, John believes that it is essential to continue this work. There is so much more to learn, and improvements in quality and efficiency must continue to be made if the industry is to keep growing and attracting new customers.

One fascinating area of research involves growing trees on dwarfing root stocks with an interstem, a complicated-sounding process which produces a tree similar to a grape vine. Supported by posts and wires, as vines are, these trees are pruned to a height of two metres and planted in a density far higher even than the trees in bush orchards. The aim is both to increase yields and to simplify the business of growing apples. Because the trees are planted in relatively low, narrow hedges, it is possible to manage them with machinery that goes up and over the whole tree. Sprays can be applied direct, and any excess captured before it reaches the ground and reused. Similarly, apples can be shaken down and caught by the same machine – before they touch the ground.

Red Falstaff and other varieties have been planted

A warm summer means sweet apples – and good cider.

in this way at Thatchers in a series of trials, and prototype machinery designed by John has been developed and tried out. This kind of research is expensive and time-consuming, which makes it all the more impressive that John has devoted so much time and energy to work that was once carried out by a team of scientists. Then again there is every likelihood that John's efforts will, in the long run, produce benefits in terms of yield and efficiency that will help Thatchers to meet new and as yet unforeseen challenges.

One thing we can say with certainty is that the cider industry will not stand still. New varieties and new methods that today seem extraordinary will one day be the norm, and among UK cider companies Thatchers is taking a lead in making this happen.

THE CIDERMAKING YEAR: WINTER

Winters in Somerset are not usually harsh, but in the dark depths of January the spring can seem a long way off. In the orchard, this is the time when the trees rejuvenate, their roots pulling nutrients up into the dormant tree ready for new growth, and it is the time when the orchard manager tends to the trees, pruning and planting.

Pruning is an essential job, especially in bush orchards where trees need to be kept in the best possible health. An unpruned tree will yield less fruit than a tree that has been carefully cut back in the winter, and pruning has been a valued skill – almost an art form – since the days when Roman fruit growers left their pruning knives lying around for us to find. The aim is to focus the tree's energy on producing fruit rather than on growing, and to give the fruit access to the sunlight that will help it develop from blossom to ripe apple, so untidy growth is removed and the branches trimmed so as to encourage growth in particular directions. Tree-planting too takes place at this time of year – once an arduous job that would take weeks but now, as we have seen, a whole lot easier.

WASSAIL!

Winter is also the setting for one of the strangest and most wonderful of English rural traditions: the wassailing of the apple trees. The origins of wassail are lost in the mists of time, but the word itself comes from the Saxon 'waes hael' or 'good health'. Originally the ceremony involved a party of revellers going from house to house singing and drinking from a communal cup, and in the early twentieth century the folk song collector Cecil Sharp found this still being practised in Curry Rivel, Somerset:

> Wassail! and wassail! all over the town;
> The cup it is white and the ale it is brown;
> The cup it is made of the good old ashen tree,
> And so is the malt of the best barley.
> For it's your wassail! and it's our wassail!
> And it's joy be to you, and a jolly wassail!

Winter in the orchards.

Members of the Mendip Morris enjoying the Thatchers Wassail.

The song continues for several verses, as the singers persuade the householder to invite them in for bread and cheese washed down with 'a drop or two of cider', in return for the singers' blessing of house and occupants – an arrangement that suited all parties in more superstitious times.

At some point in the distant past the blessing of houses in this way was extended to the orchards, and during the nineteenth century farming families went out on January 17th to perform the ceremony. Why this particular date? Once upon a time this was the date of Twelfth Night, until the Gregorian Calendar was adopted in 1752 and Twelfth Night moved to its present slot. When the calendar changed people were unsure when they should wassail their orchards and, to find out, a crowd gathered in Glastonbury on the new Christmas Day to see whether the famous thorn tree (allegedly planted by Joseph of Arimathea) would flower. It didn't. Instead, the thorn flowered on the old Christmas Day, suggesting to people that they should ignore the new-fangled calendar. This they continue to do.

Traditionally wassail was a family affair. The whole family was summoned, including the sick, the old and the very young, and a procession made its way in the winter dark to the orchard. The senior member of the family – a grandfather, usually – led the way. A pail of hot cider was brought, and a plentiful supply of toast, and anything that could be used to make a noise – pots and pans, kettles and anything else that could be carried. The family surrounded the oldest tree in the orchard and sang a variation on the Wassail Song. This one will be familiar to anyone who has attended the exuberant Thatchers Wassail over the past few years:

O lily-white lily, o lily-white pin
Please to come down and let us come in!
Lily-white lily, o lily-white smock
Please to come down and pull back the lock!

Chorus:
Our wassail, jolly wassail
Joy come to our jolly wassail
How well they may bloom, how well they may bear
So we may have apples and cider next year
O master and mistress, o are you within?
Please to come down and pull back the pin

Chorus
There was an old farmer who had an old cow
But how to milk her he didn't know how
He put his old cow down in his old barn
And a little more cider won't do us no harm

Harm me boys harm, harm me boys harm,
And a little more cider won't do us no harm

Chorus
O the ringles and the jingles and the tenor of the song goes

Merrily, merrily, merrily
O the tenor of the song goes merrily.

Spoken: (shouted!)
Hatfuls, capfuls, three-bushel bagfuls,
Little heaps under the stairs.
Hip hip hooray!

With that everyone made as much noise as possible, either by banging pots and pans or by firing blunderbusses and shotguns into the tree. Some people believe that the purpose of this was to ward off evil spirits that might have made their home in the winter orchard, but it is equally possible that the noise was meant to waken the god or goddess of the orchard from their slumber and encourage new growth.

Next, according to tradition, toast was dipped in cider and placed in the branches of the oldest tree, and cider was poured on the roots. The toast in the tree served both a symbolic and a practical purpose, since the cidery treat was an offering to the good spirits of the orchard, which were personified in the robin. At the same time farmers sought to attract robins and blue tits for pest control – a wonderful example of the way in which real-life necessities become entwined with symbolic gestures.

Many superstitious beliefs were on the wane as the twentieth century began, and wassailing with them. In 1920 a local writer noted, 'With a tenacity that is surprising in the face of modern sentiment, or rather lack of sentiment, that old and once popular custom of wassailing the apple-trees on the night of January 17th still persists to some extent in the westernmost corner of Somerset.' Apparently it survived through the twentieth century in the Minehead area, with the Butchers Arms at Carhampton being the world centre of wassailing for many years. Then, in the early 1990s, the charity Common Ground introduced Apple Day as a way of encouraging interest in orchards, and out of this grew a new fascination for wassail.

WASSAIL, THATCHERS STYLE

At Thatchers wassailing has taken off in a big way since the first winter get-together was held some years ago. Following tradition, this is a family affair rather than a public event, but here we're talking the extended Thatchers family: to celebrate the previous year's successes and in anticipation of a good year to come, John and Martin invite employees, local friends, customers in the retail trade, journalists and suppliers. Some businesses would focus on customers and leave suppliers out of it, but Martin appreciates the value of having suppliers who will drop everything to help when there's a last-minute order. So it's a mixed crowd from all over the country who turn up at Myrtle Farm in January, and for beginners the evening is quite an experience.

First, everyone gets together in the marquee to sample some of the special cider blended for the occasion – with a couple of barrels at least of the

Contemplating the beginning of another cider season at wassail.

Thatchers
Cider Company

Serenellini

proper old-fashioned stuff, and one or two sweeter, fruitier blends – and then it's out to the yard. In a scene from the Middle Ages a bonfire is blazing while the Mendip Morris Men prance and leap, led by a burly Master of Ceremonies in a top hat and a Green Man dressed head to toe in the appropriate colour, with green hands and face and a hat covered in foliage and fruit.

Between them, and very much enjoying the fun, is a teenage girl in a floral tiara (Eleanor Thatcher in a recent year). After a bit of a dance it's time for the Wassail Song (to the words above) and then the MC leads the way to a nearby tree – by tradition the oldest tree in the orchard – and the age-old ceremony continues. The Wassail Queen places cidery toast in the branches and pour cider on the roots, then the band strikes up with a tremendous racket and everyone bangs and stamps and shouts.

Now warm and smiling, the crowd returns to the marquee for an excellent dinner washed down with plenty of cider, at which point they're ready to be entertained by award-winning stand-up comics and musicians of one kind and another. A band plays favourite songs for an hour or so and then it's time to go home, the orchard thoroughly wassailed and the convivial spirit of cidermaking celebrated.

Firing shotguns into the trees may seem hazardous, but is based on sound science.

WASSAIL SCIENCE

This may seem strange, given the curious goings-on just described, but there is some orchard science – or at least common sense – in wassail. When you put lovely cidery toast in the branches of a tree you're not offering it to the tree but to hungry robins and blue tits. In the spring these birds play a vital role in eating pests, and you want as many of them living and nesting in your orchard as possible. For the same reason it's a good idea to keep old, even dead, trees for birds to nest in – at least in traditional, unsprayed orchards.

In the old days wassail involved firing shotguns into the branches of the tree, and the great Edwardian nurseryman George Bunyard thought this slightly hazardous activity made sound sense, since gunshot wounds would encourage cankers to form in the bark of the tree, and cankers prevented the downward flow of sap from branch to trunk. More sap in the branches meant a more fruitful tree, he thought, though whether anyone else but him realised this we shall never know.

Recipe for Wassail Punch: 2 litres of Thatchers Heritage Cider, 1 orange, sliced, 1 lemon, sliced, 1 apple, sliced, 2 star anise, 5 cloves (or half a teaspoon of ground cloves), 2 cinnamon sticks (or half a teaspoon of ground cinnamon), 8 slices of ginger (or 1 teaspoon of ground ginger), 5 tablespoons of soft brown sugar, to taste. Simmer gently for 30 minutes but don't boil.

SPRING

Even after all these years Martin Thatcher is still bowled over by the sight of an orchard decked in apple blossom, and he's not alone. To many people, the coming of spring in cider country is one of Britain's wonders – nature and humankind working together to create a marvel. Few sights compare to the beauty and magic of trees suddenly covered in white or pink flowers, yet it is not only a visual experience. A walk in an orchard at blossomtime is as much about the scent of the flowers and the drone of bees, the feeling of warm sun or a cool breeze – and a sense of magic.

To the cidermaker, though, blossom is much more than just a thing of beauty. For the trees to produce apples the flowers need to be pollenated, and the blossom doesn't last for long. If you're unlucky, in fact, it can disappear overnight.

Nothing makes a cidermaker quite as nervous as the prospect of a late frost, which can devastate a whole orchard in one night, and in times past fruit growers came up with some ingenious ways of keeping temperatures above freezing at this crucial time of year, such as burning oil in 'smudge pots' to warm the air and keep it mobile. On one 19th century Gloucestershire fruit farm the foremen slept beside alarms connected to thermometers, which would sound when the temperature became dangerously low.

But they were probably growing fruit varieties that were not native to the area. The beauty of the Somerset cider varieties is that they have evolved over the years in the same climate, so that most blossom after the worst danger of frost is over. Thus although Somerset Redstreak is classed as an early bittersweet, it flowers in mid-season, that is in the middle of May, after all but the most freakish frosts. However, that old renegade Tremlett's Bitter indulges in rather riskier behaviour, often flowering in late April when freezing conditions are still a distinct possibility. It is still essential, as it always was, to pick the right site for a new orchard.

ENCOURAGING BUMBLEBEES

However advanced the technology of cidermaking may become, the weather remains its old unpredictable self. Meanwhile, even the most sophisticated cider mill relies for its supply of fruit on a workforce that can be neither trained nor managed. Without bees to pollinate the blossom there would be no fruit and no cider, and in managing the Thatchers orchards John goes out of his way to encourage them. Beehives have long been a feature of commercial orchards, but honeybees won't leave the hive in cold wet weather, so a chilly spring is likely to leave many trees poorly pollinated. Wild bumblebees are bigger and tougher than honeybees and can be seen out and about in much colder conditions, so the company has begun taking steps to increase their numbers in the orchards.

Bumblebees don't store great quantities of food like honeybees, so they need a ready supply of pollen

Encouraging a healthy bee population is essential for the upkeep of a thriving orchard.

and nectar to keep them going through the summer and into the autumn, when they go into hibernation. Working with the Bumblebee Conservation Trust, Thatchers sowed seven acres around the periphery of its orchards with a specially selected mixture of wildflower seed that includes red clover, birdfoot trefoil and yellow rattle – plants which have a long or late flowering season. In spring, obviously, finding pollen is no problem…

Traditionally, spring was not only the time to be looking forward to the autumn's crop, but also when the previous year's cider finished its initial fermentation. For small-scale cidermakers this is still the case (and there may be a few hearts in mouths as the new cider is tasted for the first time) but for Thatchers the ability to reduce juice to a concentrate and store it until needed means that batches of juice are constantly being fermented, all year round. In the old days the storage of cider was a major headache, as there was always a chance of something nasty getting into it as the months went by. Concentrated juice is much safer, as the increased levels of natural sugar it contains act as an effective preservative.

SUMMER

When William and Stan Thatcher ran Myrtle Farm summer was a time for drinking cider after a hard day's work; in terms of cidermaking this was a quiet period during which you might pause to inspect the apples ripening on a tree but otherwise got on with the business of fattening livestock and harvesting hay, wheat and barley.

These days the summer months are still extremely busy at Myrtle Farm, although in a different way. In warm weather more people drink cider, so you never know when demand is suddenly going to rocket; when that happens the bottling and kegging lines work at full speed to keep the supermarket shelves stocked. Meanwhile the presses too are busy, thanks to the longstanding contract to press blackcurrants for Ribena. The sheer volume of fruit arriving each day is awe-inspiring, and it has to be pressed right away; no wonder then, that the presses are given a thorough overhaul during the spring, to ensure that they're ready for action.

As we know from Stan's adage, it is in the early part of the summer that cider pressed the previous autumn reaches full maturity and is ready to drink, so another feature of the season is the cider competition. This is no modern invention. For as long as people have got together to make cider they have had the overwhelming desire to prove their cider better than anyone else's. Competition was a feature of the Bath and West Show from the year of its foundation in 1852, and since then many other annual and biannual contests have pitted the most brilliant cidermakers in the country against one another. The competition at Long Ashton was a highlight of the Somerset summer for much of the 20th century, while the 21st has seen record numbers of entries in a plethora of contests.

Worthy Farm comes to Myrtle Farm: Michael Eavis joins Martin in the orchard.

Well, what else would you want to drink at Glastonbury Festival?

Martin enjoys the Festival spirit.

Touring the orchards of Myrtle Farm on Open Day.

JUDGING CIDER

One of the beauties of the cider competition is that it is open to all-comers. Producers in their very first year of cidermaking have won first prize at the Bath and West, although this is the exception. In the main, the winners are artisan producers who lavish the utmost care on their ciders, and as a cidermaker becomes more successful and ups production, it is always interesting to see how they fare in an environment where marketing and packaging play no part.

On a personal note, I had the pleasure of judging single variety ciders at the Bath and West a few years ago. The day was warm, the tent quite stuffy, and the ciders were laid out in bottles on a long table; there were a lot of them, each with a label that bore only a number. My fellow judges and I went from bottle to bottle, tasting a small glass from each one, and it was soon clear that a handful were much better than the rest. These ciders were not in the ideal condition for drinking, being fairly warm, and some of the weaker ones suffered as a consequence. In the end we whittled the competition down from perhaps fifty to three, and then came the task of placing them in order. As a small-scale cidermaker myself I was secretly hoping to pick some unknown producer who made cider in their garden shed, but our unanimous choice for first prize turned out to be none other than Thatchers Dabinett. Warm as it was, this cider was refreshing and moreish, with the flavour, body and mouth-feel coming through strongly.

This win was doubly deserved, considering how much time John has devoted to the Bath and West over the years, judging on a number of occasions and in recent years sponsoring a new award for the best commercial cider orchard of standard or bush trees. John is also a past winner of the Bath and West's highest individual honour, the Royal Bath and West Society's Gold Medal, recognising his lifetime's dedication and enormous commitment to the cider and apple growing industry.

OPEN DAY AT MYRTLE FARM

The scale of production may have increased massively over the years but the cider business is still one of the most convivial. Thatchers values both the cider community, as represented officially by the NACM and unofficially by networks of like-minded producers around Somerset and beyond, and the local community. The Thatcher family has been part of village life in Sandford for more than a century, as neighbour, employer and cider provider to a population that has changed a great deal over the years. In William's day most local people would have been involved in farming or local industries, and this was still the case until the 1950s or thereabouts, when people began commuting in ever-greater numbers to Bristol from the surrounding countryside, while others chose to retire to the country.

As the village and surrounding area have changed, so Thatchers has found new ways to maintain a

Thatchers Open Day, with the Mendip Morris Men.

connection with local people. And it is important, whatever may happen in the wider world, for the company to enjoy the patronage, support and friendship of those living closest to Myrtle Farm. Sponsorship plays a significant role in this, whether you're talking about major regional teams or the local Winscombe cricket and bowling clubs. The local Rotary Club has been important to John for many years, and the relationship continues with Thatchers hosting events in support of youth and community groups in the area.

Every other September, meanwhile, as the summer draws to an end and the presses gear up for action, Myrtle Farm opens its doors to anyone who wants to come in and look behind the scenes. Since the first Open Day was held at the turn of the Millennium people have turned up time and again in droves, to enjoy guided tours of the orchards and mill, to sample ciders new and old, and to chat with the cidermakers. From a few hundred curious villagers attendance has increased to several thousand, with people travelling not only from Bristol and Taunton but from as far away as Birmingham.

For local residents this is a great opportunity to see what's going on, to marvel at the new technology on show and perhaps to reflect on the fact that those giant oak vats are still there, at the heart of the operation. But the company also has a lot to gain, since an Open Day allows staff to show people that they have no secrets (other than the cider recipes!) and no mysteries. Thatchers makes cider, and although the equipment may look increasingly as though it belongs on the set of Star Wars it is only a bigger and more complicated version of Stan's old press and hogsheads.

Talking of presses, summer is almost over, and it's time to put those giant Bucher machines to work…

AUTUMN

You can store juice. You can run the fermentation vessels year-round. You can ship cider wherever you like, whenever you like. But apples have to be harvested when they're ripe and ready, which means autumn is still cider time.

HARVESTING THEN AND NOW

The apple harvest used to be a makeshift business. Apples were simply allowed to ripen and fall to the ground, where they were left until the press was ready for them, at which point they were gathered up, loaded into carts and taken away. Travellers during a wet autumn on the Somerset Levels would sometimes see a whole orchard's worth of apples bobbing about in a foot of floodwater, but in the main the apples lay quietly in the grass. Martin and his brother were not the first boys to be sent out in early December to pick up apples, and plenty are still collected in this way around Somerset today.

Some old-fashioned cider enthusiasts argue that fruit should be left to mature as long as possible, but this approach ignores a host of disadvantages, from the difficulties of picking up apples in rough pasture (especially, as small-scale cidermakers will appreciate, pasture that has recently hosted cattle) to the problems of microbial infection in the fruit. The first advance, made many years ago, was to shake apples down from the trees as soon as they were ripe, then bag them up and leave the sacks propped up against the tree trunks for collection at a later date. Anyone who has attempted to shake apples out of a standard tree will agree that this is no easy task, involving as it does vigorous work with a long stick that has a hook at one end, known as a panking pole.

More recently, tractors fitted with a mechanical arm or belt were used to shake apples down from standard trees, while in bush orchards mechanical harvesting has been the norm for decades. Apples are shaken from the trees and then gathered up by machines equipped with rotating brushes that don't damage the fruit and John Thatcher's new harvester collects the fruit without the apples even touching the ground. From there it is a short journey to the trailer and thence to the apple pits at Myrtle Farm, so that the whole journey from tree to press may take only a matter of hours. In this way the opportunities for damage to the apples is kept to the absolute minimum, and the cidermakers are given the best possible raw materials to work with.

PRESSING TIME

Apples are also arriving from orchards around Somerset, as well as from the fertile fruit-growing regions of Herefordshire and Kent. The days of farmers turning up with a sack or two are long gone; today apples arrive by lorry and are tipped straight into the pits. Not that this means the quality is any less, rather the opposite in fact. Every shipment is inspected for quality, and the growers receive a bonus from Thatchers if the fruit is in tip-top condition. As with suppliers of all kinds, Martin values relationships with orchard owners very highly, making long-term commitments to buy their fruit so that they can invest in their orchards, and inviting them to Sandford from time to time to see how that fruit is used.

With both sides committed to a mutually beneficial relationship, Thatchers can approach each new pressing season in the security that the company's growers will strive to deliver apples in peak condition, right on time. This last factor is crucial, since the autumn is carefully scheduled so that the presses can be used as efficiently as possible. While the maturing cider is allowed as much time as it needs and is never sent out into the world until it is absolutely ready, pressing has to be done quickly, while the apples are perfectly fresh.

It's worth thinking back to Stan Thatcher's day, to see just how far the business of extracting juice from apples has come. In the early 1950s he was pressing a couple of tons a day, a volume which tripled with the

As the summer draws to a close and autumn arrives, the cider apples are ready.

THATCHERS

ROTHDEAN

THATCHERS

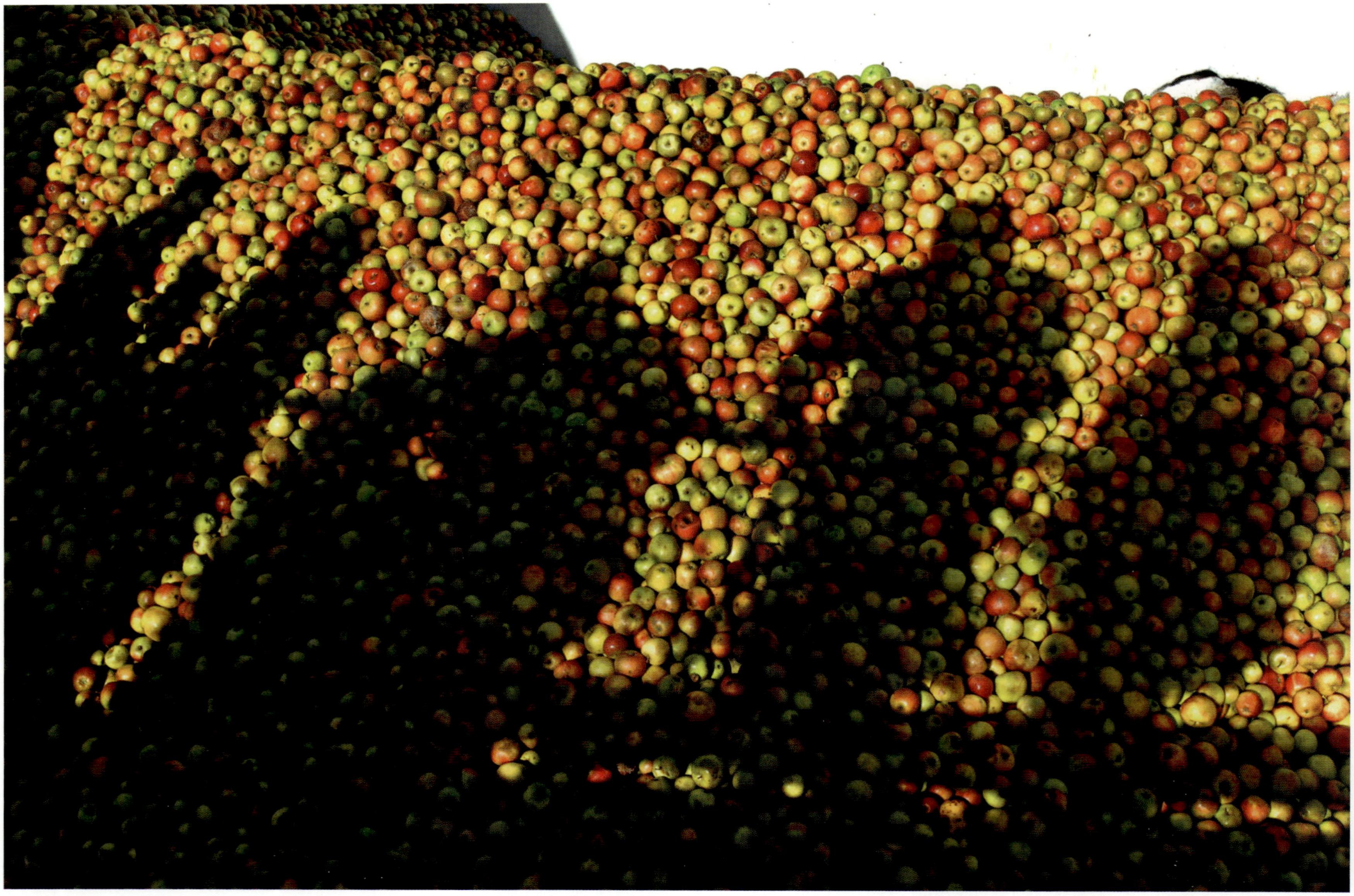

What it's all about: apples and people.

arrival of the first hydraulic press. When Martin was a boy the introduction of three new presses increased capacity to forty tons a day, with five men working all hours. Today, after decades of investment, five people can press five hundred tons a day.

From here the juice either goes straight into the fermentation vessels or into storage, ready to be pressed later. You can imagine the intensity of the work at this time, with thousands of litres of juice, from different apple varieties that need to be kept separate, being piped this way and that as yet more apples are milled and pressed, and lorries draw up one after another to tip their fruit into pits that are never full and never empty. From the beginning of September to the end of November this intense activity continues, with peaks coming as the crops of important varieties like Katy, Somerset Redstreak and Dabinett come in.

VARIETIES AND VARIATIONS

Incidentally this brings us to another important factor governing the choice of apple varieties – and the continuing quest for new ones. In the past it didn't matter too much when apples ripened and fell from the trees because it was acceptable to leave them lying on the ground, but these days, with quality at a premium and expensive equipment needing to be run as efficiently as possible, it's vital that Thatchers uses apples that are not only complementary for blending, or suitable as single varieties, but which also ripen at different times. This is not as straightforward as it seems, since the industry favourites, Michelin and Dabinett, both ripen in October, but here Thatchers' choices of apples are canny, since Katy, Cox, Tremlett's Bitter and Somerset Redstreak are ready for pressing in September or the beginning of October. Harry Masters Jersey, meanwhile, comes in a little after Dabinett, so there are apples coming through in prime condition right the way through the season. Thanks to the efforts of Liz Copas and John, the future promises more ripe apples for September, and further opportunities to increase production.

Some apples, incidentally, are easy to press, which is one reason why they have long been favourites, while others are not. Tremlett's Bitter will give up its juice if you squeeze it in your hand, but Cox's Orange Pippin offers a challenge to even the most powerful machines. No wonder cidermakers of the past used to leave fruit on the ground or in the loft for as long as possible; they may have lost a few apples here and there but those that survived were not just sweeter but also softer, and so much easier to mill and press.

So different apples have many different qualities, but there's also immense variation year on year, and one of the first jobs for the cidermaking team is to analyse the levels of sugar, tannin and acid in the juice. You can make a rough guess of how sweet a batch will be based on how much sun the apple trees have enjoyed during the growing season, but when John Thatcher was a young man there was an easy way to measure the sugar content of the juice: you'd look at the hairs on your arm after pressing and if they were all stuck together you knew there was plenty of sweetness there!

Nowadays every self-respecting amateur cidermaker measures the specific gravity of the juice and does the sums to work out what the alcohol level is going to be, and something similar, if much more sophisticated, happens at Myrtle Farm.

IN THE END, IT'S ALL ABOUT THE CIDER

Like the apples, the juice is treated as if it is precious and irreplaceable from the moment it leaves the press. So that the flavour and character of the fruit is retained, as little as possible is done to the juice; it is simply fermented, blended, matured and packaged, only on a somewhat larger scale than in the past. But everything possible is done to ensure consistency and quality, not by chucking in additives but by making sure everything is done properly, in minute detail. That's why there are temperature-controlled fermentation tanks, computer systems and a kegging line complete with robots. Like the oak vats, all this equipment has one simple purpose: to ensure that customers everywhere can enjoy Thatchers cider as much as Stan's regulars did when they shared a glass in the dovecote, all those years ago. That's what it's all about in the end: the cider.

A song and a dance in the orchard, Thatchers Open Day.

THATCHER FAMILY TREE

Benjamin Thatcher m. **Hannah**
(of Charterhouse) *(née Brunt?)*
(1760-1839) *(1770-1848)*

↓

Hester – Ann – Benjamin – John – George – Jane – Hannah – **Isaac Thatcher** m. **Harriet** – Ephraim
(of Charterhouse) *(née Hill)*
(1806-1852) *(1807-1886)*

↓

Charlotte – **Benjamin Thatcher** m. **Mary Ann** – Hannah Maria – George
(of Upper Langford) *(née Williams)*
(1841-1894) *(born 1845)*

↓

Annie – Bessie (aka Bettie) – Ernest – **William John Thatcher** m. **Beatrice**
(of Sandford) *(née Avery)*
(1880-1937) *(1883-1918)*

↓

James – Winifred – **Stanley Thatcher** m. **Christine (aka Molly) Burdge** – Reginald – Betty
(1910-1981) *(1912-2007)*

↓

John Thatcher

↓

Martin Thatcher

↓

Eleanor
Peter

Note: William John was the first member of the Thatcher family to live, farm and make cider at Myrtle Farm in Sandford
Reproduced courtesy of Maria Forbes

Apples, apples everywhere and soon a drop to drink.

John Thatcher in the orchard. There have been Thatchers at Myrtle Farm since 1904.

KEEP
CALM

THATCHERS

FAMILY CIDER MAKERS
THATCHERS
Gold
SOMERSET
CIDER